WORKING IN
JAPAN

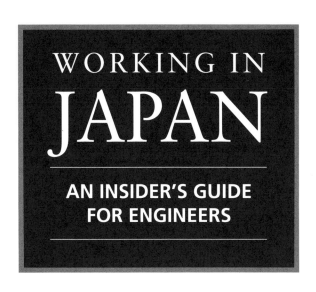

WORKING IN JAPAN

AN INSIDER'S GUIDE FOR ENGINEERS

HIROSHI HONDA

Editor

Contributing Editors

Raymond C. Vonderau

Kazuo Takaiwa

Daniel Day

Shuichi Fukuda

ASME PRESS
New York 1992

Copyright © 1992 The American Society of Mechanical Engineers
345 East 47th Street, New York, NY 10017

Second printing

ASME shall not be responsible for statements or opinions advanced in
papers or . . . printed in its publications (B7.1.3). Statement from By-
Laws.

Library of Congress Cataloging-in-Publication Data

Working in Japan : an insider's guide for engineers / Hiroshi Honda,
editor.
    p.   cm.
Includes bibliographical references and index.
ISBN 0-7918-0025-3 (pbk.)
    1. Engineering—Japan.   2. Engineers—Employment—Japan.
3. Japan—Social life and customs.   I. Honda, Hiroshi.
TA157.WG8   1992   91-26034
331.12'92'000952—dc20   CIP

Contents

Foreword

With the globalization of the engineering profession, increasing numbers of foreign-born engineers are accepting positions in Japan. Few of these engineers have the time to thoroughly prepare for their assignments, even to the extent that it is possible to do so. Dr. Hiroshi Honda and his contributing editors have prepared a book that will surely ease the transition, professionally and personally.

The busy professional will find that a wealth of information is packed into this book. The material is very interesting, highly informative, and quite frank. Its strength lies in being presented from the unique perspectives of the 19 authors. The only thing that may be missing is an explanation of the main difference between Japanese and American baseball teams. To play professional baseball in Japan, you must have *wa;* but, then, this is a prerequisite for virtually all employment in Japan.

We of the American Society of Mechanical Engineers are proud of our formal association with Japanese engineers that began in 1979 with the signing of an Agreement of Cooperation with the Japan Society of Mechanical Engineers. We are pleased that ASME Press has been chosen to publish this book, and congratulate Dr. Honda and the ASME Japan chapter for the successful and timely completion of this important project.

Arthur E. Bergles
President, ASME, 1990–91

Preface

This book is intended to help the increasing number of foreign-born professionals[1] arriving in Japan to cope with living and working in their new cultural environment. Since many of these professionals initially have a limited knowledge of Japanese culture and language, they often encounter difficulties with Japanese manners, customs, and patterns in human relations.

Foreign-born professionals are, of course, expected to bring with them some of the uniqueness and beauty of their own culture, which Japanese people do appreciate. However, there are unwritten rules which the Japanese people have developed over a long period of history for the smooth and efficient functioning of their society. If people from other backgrounds are aware of these unwritten rules and are able to adapt to them, they can enjoy a pleasant and fruitful life in Japan, both professionally and privately.

Fortunately, among the members of the Japan Chapter of ASME (the American Society of Mechanical Engineers) and their friends are people of many nationalities and cultural backgrounds; what they have in common is their experience in Japanese industry or at Japanese universities and colleges. Their cross-cultural experiences and their insights into Jap-

---

[1]The "foreign-born" engineers to whom this book is addressed include not only foreign nationals working in Japan but also Japanese citizens who were born overseas and lived there for a significant length of time. The authors of this book will use *foreign-born* to refer to both groups; however, the terms *foreign engineers* (or *professionals*) and *foreigners* will be used where more appropriate.

anese society are collected and set forth in this book, which should serve as a convenient and useful guide for the newcomer to Japanese society.

The book addresses the concerns of the foreign-born engineering professional in a logical sequence. Part 1, the Introduction, surveys the current employment of foreign-born engineers in Japan, as well as future trends revealed by statistics. Part 2, Applying for a Job in Japan, introduces effective methods of applying for a position in industry or at a university or college and describes typical salary and benefit structures in Japan. Part 3, Facets of Japanese Society, traces the history and describes the features of Japanese society, comparing it with society in western cultures. In Part 4, Cultural Gaps and the Language Barrier, some of the problems of adjustment experienced by western professionals are summarized. Part 5, The Benefits of Professional Societies, describes the opportunities afforded by involvement in the activities of the Japanese chapters of international professional societies and in those of Japanese professional societies. Part 6, Employment Case Studies, recounts the experiences of professionals of different nationalities who have worked in Japanese industry and at a Japanese university, and of some who have established their own firms in Japan. Part 7, Conclusions, first describes both the opportunities and the obstacles facing western professionals in advancing their careers after returning from employment in Japan, and then offers practical information on Japanese hiring practices based on replies given by a number of Japanese companies to an ASME questionnaire.

I hope that this book will be a valuable reference for foreign-born engineering professionals who wish to come to Japan to work.

Hiroshi Honda

Acknowledgments

This volume would not have been possible without the dedication of the authors and the support of Caryl Dreiblatt, editor, Janet Weinrib, director, Dr. Kenneth Metzner, former director, and their staff at ASME Press. In addition, we wish to thank Dr. Wataru Nakayama, chairman of the ASME Japan Chapter, Dr. Robert M. Deiters, professor at Sophia University in Tokyo, and Dr. Arthur E. Bergles, ASME president, for their advice and encouragement during the course of compiling this book. Dr. Peter E. D. Morgan at Rockwell International Science Center, Stephen Hann and other members of the ASME Japan Chapter, and many others at Mitsui Engineering and Shipbuilding Company, Beloit Corporation, ITK Inc., Osaka University, Day Translation Service, ASME headquarters, and other institutions, also provided assistance in several areas. We gratefully acknowledge all this help, without which this book would not exist.

Hiroshi Honda
Raymond C. Vonderau
Kazuo Takaiwa
Daniel K. Day
Shuichi Fukuda

PART

1

Introduction

1

Current Employment of Foreign-Born Engineers in Japan

Hiroshi Honda

INDUSTRY

As industrial operations have become more international and national economies have become more closely bound to a global economy, more and more Japanese companies and institutions have come to hire professionals from all over the world. Employment of foreign professionals, however, varies extensively among industrial sectors, and can be seen in Table 1.1. During the period from April 1989 through March 1990, 68.8 percent of Japanese securities corporations hired foreigners who had completed college or graduate school, while only 4.2 percent of energy companies hired people with similar qualifications. It can also be seen that high percentages of companies in those sectors having a high demand for engineers, such as electronics, communications, information, machinery, and automobiles, hired foreigners. As Table 1.2 shows, the communications, electronics, and automobile sectors ranked highest in average number of foreigners hired per company, while the electronics sector hired the greatest actual number of foreigners during the period. Not only are many of these companies inviting young foreign engineers to join their organizations, but also Japan's recent success in high-technology products and R&D is attracting foreign-born applicants.

A case of particular note is Mitsubishi Electric Corporation in Tokyo, which in September 1990 hired 11 foreigners who had completed college or graduate school.[1] The recruits, ten men and one woman, were from the

---

[1] *Nikkei Sangyo Newspaper,* September 12, 1990.

TABLE 1.1 Rate of Hiring of Foreigners (with College or Graduate School Degrees) by Japanese Industrial Sector during the Period April 1989 through March 1990

Sector	Number of Companies Surveyed That Hired Foreigners	Percentage of Total Surveyed Companies in This Field That Hired Foreigners
Securities	22	68.8
Electronics	44	44.4
Communications	14	43.8
Commodity trading (shosha)	9	39.1
Information	19	38.8
Machinery (factory automation)	49	32.9
Machinery (shipbuilding)	12	29.3
Department stores and supermarkets	23	26.7
Automobiles	18	26.1
Banking and insurance	22	24.7
Textiles, paper, and miscellaneous consumable products	22	20.6
Metals	9	20.0
Food and pharmaceuticals	22	19.5
Chemicals	12	14.1
Transportation and travel and leisure services	15	12.7
Home construction and real estate	22	11.5
Energy	1	4.2

Source: Nikkei Newspaper, August 3, 1990. Figures are based on hiring by companies' head offices only and do not take into account the number of foreign professionals who immediately transferred to local offices, to factories, or to an R&D center.

Cambridge University, Princeton University, and MIT, among others. All of them were engineers in high-tech fields such as computer science, electronics, electrical engineering, and systems control. Mitsubishi's intention was to send them back to their home countries to work as managers for the company's subsidiaries after they gained several years of experience in

TABLE 1.2 Number of Foreign Employees Hired (with College or Graduate School Degrees) by Japanese Industrial Sector during the Period April 1989 through March 1990

Sector	Total Number of Employees Hired	Average Number per Company
Communications	68	6.8
Electronics	150	4.8
Automobiles	73	4.3
Department stores and supermarkets	60	3.8
Commodity trading	26	3.7
Securities	70	3.7
Banking and insurance	58	3.4
Home construction and real estate	67	3.4
Textiles, paper, and miscellaneous consumable products	40	2.9
Information	36	2.8
Machinery (shipbuilding)	27	2.7
Machinery (factory automation)	94	2.5
Metals	15	2.5
Chemicals	21	1.9
Food and pharmaceuticals	25	1.8
Transportation and travel and leisure services	14	1.6
Energy	1	1.0

Source: Nikkei Newspaper, August 3, 1990. Figures are based on hiring by companies' head offices only, as in Table 1.1. The companies considered in this table are only those that gave a specific number of foreign employees in reply to the *Nikkei Newspaper* questionnaire. The denominator used in calculating the average number per company in each field therefore differs from that used in determining the percentage for the corresponding field shown in Table 1.1.

Japan. However, the term of the initial employment contract was only two years, and Mitsubishi's personnel department was concerned about whether these young professionals would remain in the company any longer, since it is not uncommon for foreign professionals in Japan to change jobs every one to three years.

Mitsubishi Electric is only one of a number of Japanese multinational companies—others are Toshiba Corporation (one of the companies of the Mitsui group), Hitachi Limited, Sony Corporation, Fujitsu Corporation, Japan NCR Corporation, Toyota Motor Corporation, Honda Motor Company, Nissan Motor Company, Asahi Glass Company, Chiyoda Corporation, and JGC Corporation—that have hired foreign-born engineers. Even heavy industries such as Sumitomo Heavy Industries and Mitsui Engineering and Shipbuilding Company, which rely extensively on domestic production and only recently began to adopt a global strategy in corporate management, are hiring foreign-born engineers.

It is also a common practice for Japanese companies to hire young professionals from the host country to work for an overseas subsidiary and then send the more promising among them to Japan to educate them in the Japanese way of corporate operation and management, subsequently sending them back to the subsidiary company. In this age of central ownership combined with local decision making, this type of employment will become increasingly common and should be well accepted by the public.

NATIONAL INSTITUTES

It is not only industry that attracts young foreign-born engineers and scientists. Japanese national institutes such as those in Tsukuba have become famous for their accomplishments in high-tech research and so have attracted an influx of foreign-born researchers. The Japanese national government has begun preaching the importance of "technoglobalism," which entails inviting foreign-born researchers to work at Japanese institutes and inviting overseas enterprises to join in national R&D projects. This will inevitably induce more and more foreign-born researchers to come to Japanese research institutes.

At the time of this writing, over 2000 foreign-born researchers, including students and research trainees, had come to Tsukuba, from 106 countries. Sixty-three percent of them came from Asia, 11 percent from North America, 10 percent from South America, 8 percent from Europe. In terms of nationality, Chinese constitute the greatest number and Koreans and Americans follow. Most of them are in their thirties.[2]

These foreign-born researchers have included a Hungarian who was most impressed with Japanese advances in enzymology research compared with research in the west and has recommended that his junior colleagues come to Japan rather than the United States or the United Kingdom, and whose family wished to stay in Tsukuba beyond the term initially planned;[3] an American who gained valuable experience in superconductivity research under the direction of Dr. Hiroshi Maeda, whose papers were the most frequently cited in the field of physics in 1989;[4] a Chinese woman who was conducting research on precision measurement using lasers and admired the level of Japanese technology;[5] a man from Finland who was conducting research on insects and appreciated the spirit of *wa* (peace, harmony, and cooperation), which the Japanese exhibit in their conduct of research and in everyday life;[6] and an American who was married to a Japanese woman and who was studying robotics and appreciated his connection with the Electrotechnical Laboratory.[7]

Many of these researchers, however, point out certain drawbacks in the work atmosphere: the traditional Japanese principle "deru kugi wa utareru" ("the nail that sticks out is hammered down," suggesting that exceptional individuality or talent might not be tolerated); the limited number of posts available for visiting foreign researchers; barriers in exchanging ideas between different sections; and intolerance of failures. But because of the successes that have come through the exchange of researchers, the government has come to realize its importance, and it is likely that the national institutes will receive more foreign researchers in the future.

UNIVERSITIES

In general, western researchers coming to Japan prefer to work at corporate research centers and national research institutes rather than uni-

---

[2] *Nikkei Newspaper Evening Issue,* August 24, 1990.
[3] *Nikkei Newspaper Evening Issue,* August 22, 1990.
[4] *Nikkei Newspaper Evening Issue,* August 23, 1990.
[5] *Nikkei Newspaper Evening Issue,* August 24, 1990.
[6] *Nikkei Newspaper Evening Issue,* August 28, 1990.
[7] *Nikkei Newspaper Evening Issue,* August 29, 1990.

versities, because of their more advanced facilities and equipment. However, Japanese universities such as the University of Tokyo, Kyoto University, and Nihon University are now improving the environment for the foreign-born researcher. In the late 1980s, Dr. Hiroshi Inose, a former dean of the College of Engineering of the University of Tokyo, spoke in favor of establishing "centers of excellence" at Japanese universities that would attract gifted native and foreign researchers. Good examples of the concept can be seen in the Research Center for Advanced Science and Technology (RCAST) of the University of Tokyo; the Institute for Fundamental Chemistry in Kyoto, directed by Dr. Ken-ichi Fukui, a Nobel laureate in chemistry and a former dean of the College of Engineering of Kyoto University; and the Nihon University institute in Tokyo, which is engaged in joint efforts with the Media Laboratory at MIT. The atmosphere, rules, systems, and facilities of the university research institutes are gradually being improved to be more appealing to foreign-born researchers and students. It must be noted that over half of all graduate students in doctoral courses at the engineering faculties of the seven major national universities are foreign nationals and that the number of these students seems likely to increase in the future.

2

Future Trends in Employment of Foreign Professionals

Hiroshi Honda

Employment of foreign-born engineers will probably increase in the future, as Japanese companies adopt a global strategy in corporate management. The trend can be seen in the figures in Table 2.1. The increases in the rate of employment of foreigners in 1991 over that in 1990 (as projected at the time of this writing) are listed by sector order of magnitude. The top eight sectors—chemicals, electronics, automobiles, communications, information, machinery (shipbuilding), machinery (factory automation) and food and pharmaceuticals—are industries having a high demand for engineers. The increase in the rate for these sectors ranges from 41.7 percent on down to 22.7 percent. The growth is likely to continue into the twenty-first century, as Japanese companies become increasingly globalized and the supply of Japanese engineers becomes more inadequate to meet the need.

So far, many foreign professionals have stayed with their first Japanese company for only one to three years and then moved to another company, partly because they wanted experience at different important corporations and partly because they felt that their contributions were not well reflected in their salaries. (It must be noted that foreigners often regard their experience at a Japanese major corporation as a stepping stone along a career path). Westerners have also complained that their job descriptions, responsibilities, and obligations are ambiguous, and that their work is not rationally evaluated. Many personnel departments of Japanese companies have taken these objections to heart and are taking measures to make foreign professionals feel more comfortable and willing to stay with their

TABLE 2.1   Increase in the Rate of Hiring of Foreigners (with College or Graduate School Degrees) in Fiscal Year 1991 (Projected) over That in Fiscal Year 1990, by Industrial Sector

Sector	Percentage Increase
Chemicals	41.7
Electronics	40.9
Automobiles	38.9
Communications	35.7
Information	26.3
Machinery (shipbuilding)	25.0
Machinery (factory automation)	24.5
Food and pharmaceuticals	22.7
Department stores and supermarkets	21.7
Transportation and travel and leisure services	20.0
Home construction and real estate	18.2
Banking and insurance	13.6
Textiles, paper, and miscellaneous consumable products	13.6
Securities	13.6
Metals	11.1
Commodity trading	0.0
Energy	0.0

Fiscal Year 1991: April 1991 through March 1992.

Fiscal Year 1990: April 1990 through March 1991.

Source: Nikkei Newspaper, August 3, 1990. Figures apply to hiring by companies' head offices only, as explained in Table 1.1, note.

companies longer. Some Japanese companies, such as Toshiba Corporation, have begun to clearly articulate a new corporate management philosophy in such areas as codes of ethics, labor management policies, and global rules in order to fall into line with internationally accepted western standards. However, problems remain with regard to salaries, since Japanese companies must balance the relatively high salaries prevailing in the world engineering labor market with the benefits they provide to their

Japanese employees. A particular difficulty is that western professionals expect rapidly escalating pay scales, even at low levels of seniority.

Mitsubishi Metals Corporation (now renamed Mitsubishi Materials Corporation), Fujitsu Limited, NEC, and Honda Motor Company, among others, recently announced plans to welcome more foreign-born engineers to positions at their offices, R&D centers, and factories in the near future. Fujitsu announced that it would increase the percentage of foreign professionals in its Overseas Business Division to 10 percent (50 to 60 persons) by 1992 from the current 5 percent. Honda announced that it would hire 170 American engineers and researchers by 1992 and an additional 300 Americans by 1995 at Honda R&D North America; as a result there will be more chances for Americans to work in Japan. This technoglobalism will also extend to Japanese national institutes and universities in varying degrees, resulting in the direct recruitment of more foreign-born engineers in Japan. At present, Japan sends twice as many researchers and engineers as it receives; as the trends described in this chapter continue, that imbalance will be redressed.

Applying for a Job in Japan

3

Looking into Companies

Hiroshi Honda and Raymond C. Vonderau

GENERAL CONSIDERATIONS

Many Japanese companies have clearly indicated their willingness to hire foreign-born engineers, as explained in Part 1. At this time, it does appear relatively easy for a skilled and/or talented foreign-born engineer to find a job in Japanese industry. But more is involved than professional qualifications; Japanese companies are also looking for candidates with personalities that they judge will be well accepted by their Japanese employees. Personality often plays a vital role in organizations in Japanese culture. Japanese companies emphasize group effort and team spirit. This is especially true of design, sales, and corporate administration departments; individualism is tolerated to a greater extent at laboratories and research facilities. Another concern of personnel departments is term of employment: though the job hopping prevalent overseas is well known in Japan and is becoming increasingly common among the younger generation, Japanese employers by and large still hope that any qualified foreign engineers they hire will, like their typical Japanese colleagues, plan for a fairly long-term commitment—at least three years, if not more.

In this chapter we will focus on how one can find a desirable position in Japanese industry. It must be noted that a position offered by a company may not initially seem attractive to a foreign-born professional. Even so, the employee will have a chance to move to another position or to work on another job if he[1] shows competency at his originally assigned job. The candidate must bear in mind that the company's motivation, just like that of companies at home, is to maximize output; a Japanese company, in hiring foreigners, has the added desire to achieve a positive influence on its Japanese employees. Therefore, the candidate's career plan must be compatible with the company's hiring policy.

EFFECTIVE METHODS OF APPLICATION

A foreign engineer seeking a job in Japan with a Japanese company should know what method of hiring is preferred in Japan. Seldom would an applicant's letter or application sent directly to a company in Japan be successful in establishing the necessary dialogue leading to employment. Japanese society relies heavily on introductions from known associates before a "new face" is invited to join an existing group or activity; this rule applies as much to obtaining an engineering position in a Japanese company as to any other activity.

The best first step is to obtain employment with a Japanese affiliate company located in one's "home" country. Experience and a good job performance record with a Japanese affiliate company are important assets, and employment with the affiliate company provides an opportunity for the "home" company managers to become acquainted with the engineer's abilities and career objectives. They can then make introductions and recommendations to specific managers in the company's Japan offices.

For a recent graduate, the college or university placement office can often provide a list of Japanese companies that have affiliates located in the applicant's home country. It is particularly helpful if a professor who is familiar with the applicant's academic record has previously recommended candidates who were subsequently hired by the affiliate company. A professor who has done consulting work for a company in Japan or the affiliate of the Japanese company would also be in a position to help. Again,

---

[1] We will use the masculine pronoun here and throughout this chapter to refer to the general candidate; no exclusion of female engineers is implied. Many Japanese women are now working in the engineering field, and foreign-born female engineers can be confident of a welcome equal to that given to male engineers.

the personal contact and recommendation are most helpful in ensuring consideration for employment.

Some engineers have also become acquainted with Japanese companies by attending trade fairs organized for the purpose of interviewing seniors or recent engineering graduates having specific kinds of experience and an interest in working in a foreign country. This means of meeting qualified engineers is often used by Japanese companies that have no affiliate company outside Japan.

If none of these routes is open, one should write for advice either directly to companies' personnel departments in Japan or to an alumnus of one's school working in Japan. Here, expert knowledge or experience will be an advantage. Since Japan is facing a shortage of engineers and scientists, especially in high-tech fields such as electronics and advanced materials, expertise in these will make an applicant very attractive. A knowledge of Japanese will also be an asset.

Besides obtaining an engineering position with a Japanese company, the engineer, making the commitment to move to Japan to work, needs to understand, at least in general terms, the following important matters relating to job satisfaction and performance discussed throughout this book:

1. Job level and promotions
2. Compensation practices
3. Importance of supporting the group effort

COMPENSATION

Tables 3.1, 3.2, and 3.3 are good references for foreign-born professionals regarding representative salaries. Japanese companies have encouraged the custom of lifetime employment through a system of awarding step-by-step raises on the basis of seniority. However, this tradition is beginning to lose its hold because of the increasing amount of job hopping that exists among young Japanese and among foreign-born professionals—a result of the globalization of the job market and industry. The conventional step-by-step raise is still valid for Japanese employees other than managers; some average figures are shown in Table 3.1. At about the age of 55, the average model salary reaches a maximum; it usually decreases afterward if the

TABLE 3.1 Average Model* Annual Salary by Industrial Sector (College Graduates, for Companies with 1000 or More Employees)

Industry Sector	Age 22	Age 25	Age 27	Age 30	Age 35	Age 40	Age 45	Age 50	Age 55
Construction (average):									
$ (U.S.)	15,880	25,360	29,480	33,920	42,820	52,930	63,310	74,141	83,100
1000 yen	2,144	3,423	3,980	4,579	5,781	7,145	8,547	10,009	11,218
Production (average):									
$ (U.S.)	16,270	24,700	28,880	33,280	42,510	52,640	64,230	77,289	84,630
1000 yen	2,196	3,335	3,899	4,493	5,739	7,107	8,671	10,434	11,425
Food and drink:									
$ (U.S.)	15,990	25,000	30,000	34,310	44,120	57,090	71,690	84,200	89,526
1000 yen	2,159	3,375	4,050	4,632	5,956	7,707	9,678	11,367	12,086
Textile:									
$ (U.S.)	16,150	24,900	29,700	33,930	43,280	55,410	68,040	77,667	78,259
1000 yen	2,180	3,361	4,009	4,580	5,843	7,481	9,186	10,485	10,565
Chemicals:									
$ (U.S.)	16,600	26,290	31,420	36,970	47,970	58,190	69,470	85,319	91,763
1000 yen	2,241	3,549	4,242	4,991	6,476	7,856	9,378	11,518	12,388

Ceramics and cement:									
$ (U.S.)	16,190	23,750	27,130	31,300	40,040	50,820	64,400	75,519	91,541
1000 yen	2,185	3,206	3,663	4,225	5,406	6,861	8,694	10,195	12,358
Steel:									
$ (U.S.)	16,270	25,070	28,920	34,310	43,930	53,570	57,640	81,037	—
1000 yen	2,197	3,385	3,904	4,632	5,931	7,232	7,781	10,940	—
General machinery:									
$ (U.S.)	16,130	23,470	26,810	30,350	36,770	48,920	61,450	68,130	77,970
1000 yen	2,178	3,168	3,620	4,097	4,964	6,604	8,296	9,198	10,526
Precision machinery:									
$ (U.S.)	16,840	25,680	31,390	36,360	47,060	54,890	81,711	94,504	91,556
1000 yen	2,273	3,467	4,237	4,908	6,353	7,410	11,031	12,758	12,360
Electrical:									
$ (U.S.)	16,670	25,440	28,810	33,020	40,810	49,040	54,530	67,260	—
1000 yen	2,250	3,434	3,890	4,458	5,510	6,621	7,361	9,080	—
Transportation machinery:									
$ (U.S.)	16,130	28,820	26,760	30,140	37,480	46,580	53,070	69,010	70,480
1000 yen	2,177	3,216	3,613	4,069	5,060	6,288	7,164	9,317	9,515

Average model annual salary = standard monthly salary (F.Y. 1990) × 12 + winter bonus (1989) + summer bonus (1990); conversion rate $1 (U.S.) = 135 yen

Model is the term used in the source; *nominal* would be close in meaning.

Source: Seisansei Model, Sougou Chingin Jittai Chosa, Nihon Seisansei Honbu (Productivity Model, Total Salary Investigation, by Japan Productivity Center).

TABLE 3.2 Average Model Annual Salary for Managers (College Graduates)

	Number of Employees per Company			
	Over 5000	1000–4999	300–999	Fewer than 299
General manager (bucho):				
$ (U.S.)	90,590	76,150	65,600	58,500
1000 yen	12,230	10,280	8,850	7,900
Manager (kacho):				
$ (U.S.)	68,100	60,000	50,500	44,500
1000 yen	9,190	8,100	6,820	6,010
Assistant manager (kakaricho):				
$ (U.S.)	53,600	44,700	38,300	34,000
1000 yen	7,230	6,040	5,170	4,590

Average model annual salary = standard monthly salary (F.Y. 1990) × 12 + winter bonus (1989) + summer bonus (1990)

Conversion rate $1 (U.S.) = 135 yen

Source: Seisansei Model, Sougou Chingin Jittai Chosa, Nihon Seisansei Honbu (Productivity Model, Total Salary Investigation, by Japan Productivity Center).

person is not a *yakuin* (director, managing director, vice president, president, or other executive). The model salary includes income and local taxes. In the case of foreign professionals, the local tax is applied only to those who have lived in Japan for one year or longer. Japan concluded tax treaties with a number of nations; therefore, foreign professionals may be exempted from their home country's taxes or they may need to pay the difference between the Japanese taxes and their home country's corresponding taxes to the home country. The model salary does not include the payment for the overtime work that young Japanese engineers regularly do at the direction of their supervisors. If the payment for overtime work is included, the actual annual salary will range from 110 to 130 percent of the model salary, depending on the company. The annual salary may reach almost twice as much as the model salary if the employee is extremely busy.

Table 3.2 shows average model salaries of general managers (*bucho*),

TABLE 3.3   Percentage Index of Average Model Annual Salary by Age of Employee and Size of Company (College Graduates)

Number of Employees per Company	Age 22	Age 25	Age 27	Age 30	Age 35	Age 40	Age 45	Age 50	Age 55
Over 1000	100	100	100	100	100	100	100	100	100
300–999	97	94	91	90	87	86	86	85	86
100–299	95	94	90	89	86	84	82	82	83
Fewer than 100	93	93	88	89	86	83	78	78	74

Average nominal annual salary = standard monthly salary (F.Y. 1990) × 12 + winter bonus (1989) + summer bonus (1990)

Source: Seisansei Model, Sougou Chingin Jittai Chosa, Nihon Seisansei Honbu (Productivity Model, Total Salary Investigation, by Japan Productivity Center).

managers (*kacho*), and assistant managers (*kakaricho*) of companies of different sizes. In the case of bucho and kacho, who would not be members of a union, there is generally no compensation for overtime. Table 3.3 shows average model salaries according to age for different-sized companies. As the table shows, the salaries paid by larger companies are usually higher than those found in smaller companies.

It must be noted that the ratio of the yen to the U.S. dollar fluctuated within the range of 120 yen per dollar through 150 yen per dollar during the three-year period ending with the 1990 fiscal year. The salaries in U.S. dollars shown in Tables 3.1 and 3.2 are based on a rate of 135 yen per dollar and thus are accurate within a range of about plus or minus 10 percent.

The salaries of foreign-born engineers are usually set on a short-term basis (two or three years) or a project basis. A foreign professional's salary can therefore be higher than the salary of a Japanese employee having equal qualifications. Japanese companies are willing to bring into their lifetime employment systems any foreign-born engineers who are well accepted among their Japanese colleagues. However, these people usually prefer employment on a temporary basis with its higher salary levels. The salary level for the foreign professional varies extensively with the employee's qualifications. In contrast, however, Japanese college graduates of the same age are treated equally at the initial stage of employment regardless of whether they have bachelor's, master's, or doctor's degrees (and also without regard for any extra years after high school spent in preparation for college entrance examinations; about half of high school graduates fail when they first apply for admission to universities); advancement thereafter depends solely on the individual's performance and the company's policy.

Fringe benefits provided by Japanese firms usually include coverage of commuting expenses; the use of inexpensive company dormitories, with breakfast and supper provided, for single persons; inexpensive company apartments, for married persons; inexpensive (sometimes free) sports and leisure facilities; and use of resort facilities at discount rates. Reasonably priced health insurance and paid holidays are also offered. It must be noted, however, that the content of the fringe benefit package varies substantially among companies. Foreign-born professionals should therefore ask the personnel department about a company's fringe benefit system.

4

Finding a Job at a Japanese University

Shuichi Fukuda

In the Meiji era, that is, in the late nineteenth century, many of the teachers at national universities came from western countries. Similarly, far back in history, in the sixth and early seventh centuries, much of what was important in Japanese culture as well as much of Japanese technology was brought to Japan by Chinese and Koreans, who played the roles of teachers at that time.

In modern times, however, it was very difficult until quite recently for a foreigner to find a position in a Japanese university, even though many of the universities were established only after World War II. There have long been foreign-born professors teaching in Japan, but in most cases only at missionary or private universities. In contrast to the Meiji era, national and municipal universities have closed their doors to non-Japanese; those few fortunate enough to find positions were in most cases hired as assistants; it was almost impossible to become a full faculty member.

A new law enacted in 1982 has gone into effect that stipulates that foreigners may be employed in national and municipal universities under the same conditions of employment as Japanese. This law provides that a foreigner may obtain the position of professor, associate professor, or lecturer and may participate in faculty meetings with all privileges, such as voting, just as Japanese may. The term of employment is left to the individual university. In most universities, the term for a foreign faculty member is two or three years, after which the contractee has to reapply. But wages are increased in subsequent terms.

The unusual and difficult experiences foreigners are likely to have working in Japanese university laboratories are well described in the article "Maximizing Mutual Benefits for Your Stay in a Japanese Laboratory," by Professor Robert G. Latorre, in the book *Science in Japan*.[1] The article and the book provide much valuable information on how to apply for research jobs in universities, national institutes, and private industry. Those who are interested in obtaining a research position in Japan should give the book at least a cursory glance before applying.

If you wish to obtain a research position in a Japanese university, the most important thing you have to do is find a professor who shares your interest in your research plans and who is willing to take charge of your application. This is true even for Japanese who are applying for university positions, even though positions are advertised publicly and are theoretically open to all applicants. In Japan, everything is run on a person-to-person basis, so that it is vital to find a professor interested in employing you or willing to introduce you to friends and colleagues.

But you have to be very careful to take into account the Japanese way of management in a research setting, as you will understand after reading *Science in Japan*. In Japan, professors (and researchers) in principle, do not move from one university to another. Most of them work at the university from which they have graduated; therefore, you cannot simply file an application one day and begin to work the next. If you really wish to work in Japan, you must remember that not only you but also the employer will have to make preparations. This is the main difference between universities in the western world and those in Japan. Every decision is made on the assumption of a lifelong (or at least long-term) commitment. Thus, if the university you wish to work for is not interested in your research subject, you will have to either persuade people there to take interest in it or find another theme. In the former case, you will have to make a commitment to stay for a long time; if you then plan to move to another university after a short stay, you will have to be prudent because of the potential problem. It is customary for a Japanese professor to take the initiative in finding a replacement.

These principles are fundamentally the same for national, municipal, and private universities. But private universities are subject to far less strict

---

[1]Robert S. Cutler (ed.), *Science in Japan: Japanese Laboratories Open to U.S. Researchers*, Technology Transfer Society, Indianapolis, 1989.

laws regarding the employment of foreigners. They can hire anyone they can afford if they judge him or her worthy. The same flexibility applies for Japanese applicants, too. Private universities are also more flexible as far as the research environment is concerned. But private universities have their disadvantages, too. In most cases there are more students in private universities, so that more time has to be spent on teaching than in research. In national and municipal universities, annual budgets are determined by law and very difficult to modify; thus, the numbers of professors and students are limited. Therefore, generally you will have more time for research at national universities. Yet this budgetary situation in the public universities limits flexibility in employment of foreigners and the selection of research subjects. You cannot make a research proposal in Japan as freely as you can in a western country. And even if you could, you would be unable to hire a research assistant from the project money. Researchers who share laboratories in Japan share equipment, assistants, advice, and even glory to some extent. Thus, you will have to convince the other workers in your laboratory of the worth of your project, or else you will have to modify your objectives, settling on a less ambitious theme or one that fits more closely into the existing operations of the laboratory.

I cannot overemphasize how essential it is to make sure that the laboratory situation at your chosen university is compatible with your objectives, because once you have begun working at one university laboratory in Japan, you are likely to find being admitted to another a prohibitively time-consuming task requiring delicate manuevering. If you come to Japan with the express purpose of working at more than one laboratory, you should make this clear at the very beginning. Your advisers will then try to find positions which best fit your situation. In all probability these would be lecturer's or professorial posts, where you would have the rights guaranteed under the 1982 law and a salary typical of that paid to foreign-born faculty members. In such a post, however, you will be allowed more flexible management of your laboratory and, therefore, closer adherence to your research interests. But such positions are few and hard to get.

To summarize, it is very important that you maintain an attitude of cooperation both when you apply for a position and afterward when you begin working. Even if the position you are offered seems to be a little different from the one you want, take the opportunity and work within the existing framework at the new laboratory. This is the Japanese approach to research; it will allow your colleagues to feel most comfortable with you. Of course, you should be firm about your opinions, but you should not stick with them to the last. You should listen to all objections, assess the situation, and try to solve problems together with the other researchers in the laboratory. And remember that in Japan, you are expected to create the situation you want where you are, rather than look for it somewhere else.

PART

3

Facets of
Japanese
Society

5

Overview of Japanese Society

Hiroshi Honda

COMPARISON WITH OVERSEAS COUNTRIES

Japan is a small island country adjacent to the Asian continent with a history extending back about 2000 years. Japan has certain characteristics in common with Great Britain. Both the British and the Japanese have a reputation for attaching great importance to common sense; also, Great Britain, like Japan, is a small island country adjacent to a continent. Japan has been influenced by China, Korea, and India on the Asian mainland, and also by Europe and the United States and has assimilated much from cultures all over the world. Similarly, Great Britain has been strongly influenced by the European continent and has also borrowed from many other cultures. During the process, the Japanese adopted many customs, ideas, and beliefs from other countries. For example, the Japanese enjoy French, Italian, Chinese, American, German, Korean, and Indian food, among others, in addition to Japanese food. Buddhism, Shintoism, and some sects of Christianity are common in Japan. Yet, in spite of their eclectic tastes, the Japanese are obviously different in nature from the people of other nations. It is generally said in Japan that the Japanese do not seem aggressive because they tend to assert themselves indirectly and modestly in everyday life. The Japanese see themselves as rather passive, Americans more active, when the two peoples' social standards are compared. It is generally accepted that Japan is a homogeneous, conservative society descended from an agricultural people, while most western nations are made up of descendants of hunting people.

HISTORICAL BACKGROUND: FROM RECENT CENTURIES TO MODERN TIMES

In the Edo period (from 1602 to 1868), the Tokugawa feudal government adopted a policy of spreading Confucianism in Japanese society in order to maintain its power as long as possible. (Edo is the former name of Tokyo, where the governmental administration was located, away from the imperial palace in Kyoto.) This philosophy of the ideal of a peaceful and just society still holds sway in Japan, as it does over much of Asia. Under the principles of Confucianism, elderly people are always to be respected and children are to obey parents' directions and opinions. Family responsibility is also important under Confucianism. If a member of a family commits a crime for example, all family members share in the penalty to some degree. Because of this philosophy, family ties became very strong in Japan.

The *samurai* occupied the highest social rank. A samurai was obliged to serve only one sovereign and devote his life to him. Changing his allegiance meant betrayal of his society and was punished by banishment from society.

The farmer was next to the samurai in social rank. In Japanese agrarian life in particular, strong cooperation among farmers was an absolute necessity in case they needed to cope with adverse climate and weather. In times of calamity, farmers had to share the burden, and so they developed a solid team spirit. No single person, no matter how brilliant or capable, could be better off than his peers. The farmers had to share their wealth as well as their burdens. The *nanushi*, or village headman, was succeeded by his direct heir generation after generation, regardless of the heir's capability. *Gonin-gumi* (five-person teams) were formed so that every family's behavior could be observed by representatives from four other families, thus maintaining the social order. Because of this system, families had to take special care not to wrong their neighbors or even to stand out from the community, since this could lead to jealousy and envy.

An individual's shame was a family's shame and often a local community's shame. When a community was disgraced by the conduct of the samurai, he had to restore its honor by committing *harakiri* to compensate for the shame, while ordinary people were punished by authorities. (This traditional Japanese ethic was illustrated in a recent scandal, after it was revealed in 1987 that a subsidiary of Toshiba Corporation had violated the rules of COCOM, the Coordinating Committee on Export Controls. The chairman and the president of Toshiba resigned, even though the home office had not been involved in the affair). Individuals or families that wished to initiate a new enterprise could only do so with the consent of the rest of the five-man team or of the community; in addition, one family's failure was usually considered the team's failure. Denunciation and criticism were taboo, since both denouncing and denounced persons were considered disreputable. A common proverb in Japan, "silence is golden, talkativeness evil," probably stemmed from this attitude.

Human relations were thus very close, caring, and confined—certainly by American standards—as a result of the social system and ethics of the Edo period and perhaps also because of high population density in Japan. (On the positive side, the restrictive standards minimized the crime rate and maintained the social order.) The groundwork of present-day Japanese society was laid during this period—the "vertical" structure of Japanese society, as described by Chie Nakane (see Chapter 7 of this book).[1]

Tokyo: Capital of an Internationalized Japan

Also during the Edo period, international trade was prohibited, to prevent the influence of outside social systems and religions, such as Christianity, from taking hold. But in the Meiji era (from 1868 to 1912), which began after the Tokugawa feudal government was overthrown, western culture streamed into Japanese society in pace with the increase in international trade. It was at the beginning of the Meiji era that Edo was renamed Tokyo, which literally means "east capital," and became the capital of Japan. Western democracy and liberalism as well as western goods gradually penetrated into Japanese society. However, the Japanese emperor was still

[1]Chie Nakane, *Tate-shakai no Ningen Kankei* ("Human Relations in the Vertical Society," according to Robert M. Deiters' translation), Kodansha, Tokyo, 1967.

regarded as a god by all Japanese. Japan went through two industrial revolutions, like the one that had occurred earlier in the west, from the late 1800s to the early 1900s, though it was not until 1945, the end of the World War II, that democracy took root in Japanese society. It was the military occupation under Douglas MacArthur that approved the Japanese constitution and the postwar political system and reduced the status of the Japanese emperor to just a symbol of the nation.

In the years since 1980 in particular, Tokyo has become especially westernized, Americanized, and internationalized as Japan has become an economic power in the world. Changes in working life have taken place, too. In contrast with the company loyalty and lifetime employment system which were features of the period of high economic growth from the 1950s through the 1970s, job hopping has become common and the younger generation has become less interested in maintaining the strong human relations that were a traditional force in the workplace. However, the atmosphere and manners of mainstream society are still largely reminiscent of those established in the Edo period, especially in smaller cities and rural areas.

Other Important Centers of Japanese Culture

Kyoto literally means the "capital of capitals" (or "metropolis and capital") and was the capital of Japan for 1100 years until the end of the Edo period. People in Kyoto take special pride in their ancient tradition. They jokingly claim that to qualify as a "real" Kyotoite one must be descended from families living in Kyoto since before Ohnin-no-ran (a rebellion that occurred between 1467 and 1477 during the Ohnin era), whereas one can be considered a "real" Tokyoite if one's family has lived in Tokyo for only three generations or longer. Kyotoites are known for being gentle in manner but cool toward newcomers at first. However, after a substantial period of acquaintance ranging from several years to some decades, a new person can gain acceptance as a member of this traditional society.

Osaka, located only 30 kilometers from Kyoto, is a city of merchants with an open atmosphere. Very close to Osaka is Kobe, a famous seaport with an exotic, western atmosphere. Nara was a capital of Japan from 710 to 784, before Kyoto was. In the Kyoto-Osaka-Kobe-Nara region there are remains of ancient capitals from times before the Nara period. This

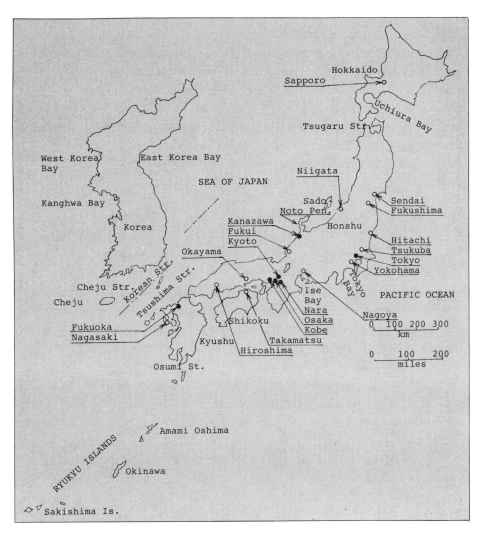

Figure 5.1 Japan and its major cities.

region is especially noted for maintaining Japanese culture and traditions and has thus attracted many visitors from overseas.

Many other Japanese cities, such as those shown in Figure 5.1, are also noteworthy as centers of Japanese culture and traditions. Kanazawa, for example, is a famous castle town with a conservative atmosphere on the side of Honshu toward the Sea of Japan; it was governed by Lord Maeda in the Edo period. People here are noted for being very polite and humble and for treating guests with special hospitality. Kenrokuen in Kanazawa is regarded as one of the three finest Japanese gardens. Fukuoka is a center of culture, commerce, and transportation on Kyushu and also a castle town that was governed by Lord Kuroda in the Edo period. The culture of Fukuoka as well as other cities of northern Kyushu has been much influenced since ancient times by the continental Asian culture, as can be easily understood from their geographical location; many of these cities were constructed by landlords in or before the Edo period.

THE MODERN WORK ENVIRONMENT IN JAPAN

The Japanese have a saying, "Capable hawks hide their talons," the meaning of which should be obvious from previous parts of this chapter.

A Japanese tends to behave in a way that is inoffensive and often complimentary to other people, not only his[2] bosses but also his junior and senior colleagues. A polite Japanese would say, "I agree with you in a way; however, there is another idea, like this . . . ," even if he actually rejects your idea. A Japanese seeks to gain the support of other people, which he hopes will lead to substantial cooperation from a majority of the members of his organization and, in turn, promotion. In contrast, promotion in an organization in western society tends to rely more on an individual boss's preference.

A member of the Japanese elite is expected to make considerable sacrifice, or at least to appear to the surrounding people to be doing so. He performs many of the tasks before him for little monetary compensation but is rewarded with the satisfaction of having the respect of his senior and junior colleagues. A respected department manager is willing to sweep the floor ahead of his subordinates, when it is time for the department members to clean the office (as they indeed do in Japan).

There is no specific job description for each individual worker, and in most cases an individual's responsibilities are not clearly defined; however, the mission of a department or office is clearly stated. Therefore, the person who contributes the most to his department or office will be promoted

[2]"His" is used here and throughout this chapter in reference to both sexes. No slight is intended to women, who are gaining slow but certain acceptance in Japanese society as individuals with potential for leadership.

first, but with only a modest pay increase. No matter how professionally capable someone is, he will not be promoted ahead of his peers unless they both recognize his effort and feel respect for him. Harmony is valued, not competition, within a group. "He is working hard, has a nice personality, and is also well accepted by his peers" is the kind of evaluation that is often heard of those who attain the greatest success. (Under these circumstances, it is also true that less capable persons tend to think that they have as much chance to be promoted as anybody, and indeed, much to their relief, they often are promoted early in their careers along with the best in their department.)

Nemawashi (literally, "root binding," or gaining a consensus of the necessary people) is also an important process in Japanese society. Nemawashi takes much time and effort. However, once it is completed, it can result in very quick action, or it can mean that a project or policy will be kept alive even if problems come up of the kind that in the west would usually result in cancellation. *Nominucation* (a slang combination of the Japanese word *nomi*, "drinking," and the English word *communication*) is a very important part of nemawashi. What was said during nominucation rarely surfaces in the more formal atmosphere of the office, but nemawashi would not function well without it. People who are good at nemawashi tend to be promoted fast in Japanese society.

When playing on a softball team at Pennsylvania State University, I noted that the batting order announced for the next game was constantly being changed by the team captain according to the performance of each player in the last game. Players who fell to low positions in the batting order seemed to take their responsibility less seriously and just enjoyed playing the game. It seemed to me that much of the way American society works fits a similar pattern. In contrast, Japanese society tends to maintain established "batting orders" until it becomes truly necessary to change them. Japanese companies hold a long-range view, of which one manifestation is the lifetime employment system. It must be noted, however, that western companies such as IBM, Du Pont, and Siemens also emphasize their own corporate culture, thoroughly educate their employees through internal training programs, and refrain where possible from laying them off.

Because of the lifetime employment system, Japanese organizations may have to set salaries lower than those of their western counterparts, since the organizations will still be subject to business ups and downs over the long run. For example, even the annual salary of a president of a major Japanese corporation is roughly in the range of $200,000 to $700,000, while salaries at the top of major American corporations range from $1 million through over $20 million, if all compensation including stock options is taken into account. It is well known that the salary of Akio Morita, cofounder and chairman of Sony Corporation, is lower than that of the president of Columbia Pictures, which was acquired by Sony.

FOREIGNERS IN JAPANESE SOCIETY

After reading the preceding sections, some westerners will feel that they will be unable to conform to Japanese society. However, non-Japanese-speaking people from other countries are treated as guests for the first few years, and many of them have an enjoyable life in Japan. During the initial period, the Japanese observe the personalities of these people; subsequently, some are accepted by the Japanese and some not. The foreigners most successful in Japan, of course, are those who have the ability to communicate in Japanese, in addition to their expertise, capability, or knowledge. Once a foreigner can communicate in Japanese, he has an invaluable tool that will gain him respect from the Japanese people, thus making it easy to be accepted. Good examples are Jesse Takamiyama and Konishiki, sumo wrestlers from Hawaii; Kent Gilbert, a television personality and lawyer from Utah; Kent Derricott, a television personality and missionary from Utah; and university professors such as Gregory Clark and Robert Deiters at Sophia University. There are also numerous western businessmen at securities corporations, investment banks, and so on, who can speak fluent Japanese and understand Japanese culture.

The Japanese are, in many instances, emotional rather than rational in making a decision. Once the Japanese feel that a foreigner truly understands and appreciates the Japanese mind and customs, they will respect him and open their minds regardless of his nationality. This is just as true in other countries. The Japanese government welcomes the contribution of truly outstanding foreign nationals, once it is judged that their personality is well accepted by Japanese people. A very good example can be seen in Sadaharu Oh, a world record holder for home runs in professional baseball, who is a Chinese national and yet won the National Honor's Award, which has been awarded to only a few Japanese.

References

The following books are recommended for readers wanting to know more about modern life in Japan:

Your Life in Tokyo, Vol. 1, *Daily Life: A Manual for Foreign Residents,* Japan Times, Ltd., Tokyo, 1987.

A. M. Newman (ed.), *Living in Japan,* 10th rev. ed., American Chamber of Commerce in Japan, Tokyo, 1987.

Industry

Kazuo Takaiwa

STAGES IN INDUSTRIAL DEVELOPMENT SINCE WORLD WAR II

It was only after the end of the Second World War in 1945 that Japanese industries began to walk alone. Before and during the war, they had faithfully delivered gigantic, ultramodern battleships, such as the Musashi and Yamato, and some of the fastest and highest-performance fighter planes, such as the Zero and Shidenkai. However, industry was the slave of the military establishment; its independence came in 1945.

The postwar years can be divided into the following general eras:

1945–1950

Immediately after the war, the Japanese people concentrated their efforts on rehabilitating their devastated land and cities with aid from the Government Account for Relief in Occupied Areas (GARIOA) fund administered by the United States. Individual survival was a struggle during these grim years.

The ravaged earth produced little, and food was constantly in short supply. Meanwhile, laborers, newly awakened to their civil rights, began to organize into unions. Labor disputes broke out often. However, in keeping with the Japanese principle of *wa* (harmony), unions and companies usually cooperated in finding solutions to the disputes. Workers did not consider it a threat when their leaders were hired into management. It was an era when production was maintained only by human power, because machine power was scarce.

1950–1955

War broke out between North and South Korea in 1950. The Japanese islands became a logistic base for the United Nations forces. At the same time, Japan was becoming ready to wake up, to stand again and walk by itself.

Japan was faced with a gap in its technical and industrial knowledge and had to dip into its still depleted purse for the funds to fill the gap with advanced technology from American and European companies, without regard for the cost. Japan has few natural resources, and the leaders of the nation understood very well that the country could not presume upon the GARIOA fund forever. Thus, Japan entered an era of technology transfer and promotion of employment. Industries invested in worker training and gradually modernized their factories, and in the process many of the small-scale factories that had dominated Japanese production since the beginning of industrialization during the Meiji era were phased out.

1955–1965

Coastal industrial zones were rapidly developed, with the construction of steel mills, oil refineries, and other petrochemical complexes. Shipyards were enlarged, eventually to scales enabling the construction of mammoth tankers. Gigantic blast furnaces were purchased and installed in more and more steel mills. Hydraulic and steam power plants were established. Automobile manufacturing swelled. Unemployment, an endemic problem since the war, went through a steady fall. Thus began the era of high growth. Industries needed foreign engineers in those heady days, but only as short-term advisers.

1965–1970

In the next phase the domestic market for industrial products and general production capacity became saturated, resulting in heightened competition among different manufacturers, engineering firms, and construction companies. Contractors began increasingly to look abroad for projects, and exports in general expanded. Japanese sales forces emphasized their strong points of rapid delivery, high quality, manufacturers' willingness to adapt to local standards, and use of electronic data processsing to improve service. This era marked the beginning of sharp competition among Japanese makers, design of products for foreign consumers, and cultural and trade friction. Industries began to employ foreign engineers under longer-term conditions.

1970–1975

A steadily increasing number of atomic plants went into operation, while pollution and other negative effects of the high level of development began to loom as a public issue. A recession occurred after the leap in oil prices. Antipollution technologies were developed, alongside a general, ongoing automation, computerization, and streamlining of industry and construction, pushed by changes in the yen-dollar exchange rate and the steadily growing price of energy.

1975–1985

Companies began to export large-scale production facilities in module form on contract for overseas buyers, and eventually Japanese companies began to shift production abroad. The emphasis in manufacturing moved to zero defects, zero accidents, and total quality control, thanks largely to the total productive maintenance (TPM) program. This was an era of large-scale production and improvements in quality. The number of foreign engineers increased, and, subsequently, the number of intercultural problems.

1985–1990

Another recession occurred after 1985, and studies began of branching into new fields such as innovative applications of information systems, biotechnology, and new materials, as well as of the potential use of new environments, such as space, the oceans, and deep underground. The expansion of overseas production facilities proceeded apace, and the employment of foreign workers in Japan, legal or illegal, became a widely discussed social phenomenon.

SOCIETIES IN CONTRAST

An old Japanese proverb says, "Don't wipe another man's face with your duster." This means, do not say things to another person that he or she is disturbed to hear, or otherwise behave in a deliberately offensive manner. Courtesy in Japan dictates avoiding saying something to someone which would put him or her in an embarrassing position, or causing other difficulties, even in the competitive world of business. The polished Japanese strives for harmony (*wa*).

Europeans and Americans on the other hand believe in claiming personal rights in debate, regardless of the potential for embarrassment or irritation of either side. This style lends itself to confrontation, something most undesirable in Japanese society. Similar western manners and ways of thinking have been propagated throughout the world for the last 400 years, while during the same period Japanese ways have been confined to Japan.

Generally, when Japanese work in close contact with foreign business-people, they try to behave according to their counterparts' culture and manners, so as to give the maximum impression of compatibility. As a matter of principle, those who wish to do business in a foreign country must follow the laws, regulations, and culture of that country. Therefore, the Japanese expect this effort of those who come to live and work in Japan. Foreign-born engineers coming to work in Japanese companies and universities should expect to find cultural differences, learn what is expected in Japan, and try to follow the rules and manners of the Japanese as far as possible.

SKETCH OF A TYPICAL JAPANESE COMPANY

Conditions of Employment

Japanese workers have no specific employment contract with their employers, except those who are working after retirement or on a temporary basis. (Recently, however, companies have begun to provide written contracts for foreign employees.) Instead, at the time of their hiring employees are requested to submit oaths with a letter from a guarantor. Companies give entrance examinations to applicants from universities or graduate schools six months prior to their graduation. Companies issue regulations which they expect everyone, including foreign employees, to follow. The company where you work may not have an English version of its regulations, which will be troublesome, but personnel department staff will explain them for you; you can always ask your colleagues if you need more concrete explanations.

Job Description

The organization chart when rendered in English will probably seem easy enough to understand at first glance, but sometimes the actual jobs employees and managers do will differ considerably from jobs in your country having the same names.

For instance, *supervisor,* in the United States, means a person who watches subordinates, guides them in performing their jobs as expected, evaluates their performance, bears responsibility for safety, and sees to it a large enough workforce is on hand to meet deadlines. Supervising subcontractors is deemed a duty of a field engineer. However, in Japan, a supervisor has more responsibilities than this. He does all the above jobs plus those of field engineers, such as subcontract relations, quality control, and progress investigation.

Laborer in the United States suggests a worker who performs unskilled physical work. Such a person does material handling, cleaning, basic setup and teardown of equipment in the field, and digging, for example—but not derusting work with wire brushes, or painting for rust prevention, for example, which are done by apprentices. In Japan, however, the "unskilled" work a "laborer" is expected to do might include derusting, painting, or sawing logs for temporary fences.

Autonomous Activities

Engineers, plant operators, clerks, and even supervisors are expected to keep machinery, platforms, field sites, and offices clean on their own as they deem necessary, without instructions from their managers. It is the duty of any employee to keep the work area clean at all times and a fundamental principle of the philosophies of total productive maintenance (TPM), the zero defect movement (ZDM), and total quality control (TQC), all of which are carried out by small autonomous groups.

Autonomy is the watchword in Japan. Employees in Japanese companies make it a point of pride to do their jobs on the strength of their own motivation without instructions from their managers. Managers often ask for opinions and suggestions from their subordinates when making certain decisions; they sound out the concerns of their subordinates on issues, and expect informed cooperation from them when the decision is finally taken. Managers do have their own opinions, but they like to follow this procedure in decision making.

Japanese are not as shy as they appear to westerners, but they respect modesty. In meetings, therefore, it is not polite to state one's opinion without permission before one's superiors have expressed theirs. This comes from the Confucian teaching of "elders first, juniors second."

The Simplest Way to Fit into Japanese Culture

A common and effective form of communication in Japan is *nemawashi*, which means laying the groundwork for obtaining the agreement of one's companions, a kind of behind-the-scenes maneuver. It is a form of communication between parties outside the normal flow of information via management. It is not always appreciated in a western organization, as it bypasses the chain of command.

A majority of those who go to foreign countries to live and work suffer frustrations. They are working within a different culture, have difficulties with communication due to the language barrier, eat unfamiliar foods, and must accept a new style of living. These shocks can be eased greatly by having the proper attitude, however, which it is the intent of this book to foster.

The foreign employee is urged to feel free to consult colleagues on all questions about job procedures, evaluations, promotions, and raises if written versions of the rules are not available in English.

Finally, I would urge all who come to Japan to make a sincere effort to master the Japanese language, since this is the most effective way to feel at home with your fellow employees and with Japanese culture in general.

I wish to express sincere gratitude to Daniel Day for his assistance with the English language during the course of writing this chapter.

7

University Education in Japan: A Personal View

Robert M. Deiters

Many readers of this book will be engineers just arriving in Japan (or only contemplating coming to Japan) to work in industry. But even for those readers who have no intention of teaching in a Japanese university, my comments in this chapter should illustrate some important points. It is only discouraging and frustrating to wail about the unfamiliarity of the customs or your inability to work in ways with which you may be uncomfortable initially.

Whether you are in an industrial or an academic situation, look first to see what cultural and human factors in your new milieu promote cooperation, a sense of belonging and participating, creativity, and growth. Use these. They are the real strength in the Japanese way of doing things.

EDUCATION IN JAPAN

Recently, higher education in Japan has drawn criticism both within Japan and from foreign observers: the curriculum is not tightly organized, observers say; little work is required of the students, and anyone can expect to graduate once he[1] gets in; and many students spend their time at side jobs, traveling, or at work on either their own personal projects or

---

[1] Throughout this chapter I will commonly use *he* and *his*, because, of the hundreds of students I have taught over the years, most were young men, with only a sprinkling of women among them. Consequently, I ask the reader to allow me to use these masculine pronouns, because I can express my actual experience more naturally in this way.

those of the numerous student clubs. Some critics go so far as to describe the Japanese university as a large—and, for the parents, expensive—"recreation center."

It is now some decades since I introduced myself in halting Japanese to my first freshman class. To some extent I have to agree with these criticisms, and—what is even more embarrassing—I have to admit that I have been part of the problem in over 30 years of teaching in Japan. But as is often the case when the works of humankind are judged, the critics have been more adept at pointing out the faults of the Japanese university than at pointing out its real strengths and potential.

In 1954 I was 29 and the members of the A-4 class (each class of 40 to 50 students becomes a social unit with a unique identification number) were 19-year-old freshmen—that was my first class at Sophia University (in Japanese, Joochi Daigaku). Just recently the "A-4 Old Friends," now in their mid-fifties, gathered. I and another former teacher were invited on their annual overnight trip to a hot spring resort. Although I had taught the class for only two hours a week in the freshman year and then had left teaching for some years to complete my own studies, they have kept friendly contact with me over the years. As a foreign teacher in Japan I sometimes experience frustrations, but the deep satisfaction of my friendship with students and graduates far outweighs the frustrations. I have some critical opinions about university education in Japan as I have experienced it, but deep down I know that a system of education that can form a group like the A-4 Old Friends is basically sound.

Beginning about the late 1970s, when the productivity of Japanese manufacturers began to draw the notice of Americans, articles and books about "Japanese-style management" came out one after another. By now, though, the more perceptive of Americans have already realized that labor-management relations in Japan are rooted in Japanese cultural and social patterns that cannot be simply transplanted to America. Improvement by imitation will not succeed.

Ever since the postwar Allied occupation of Japan changed the educational system to match—outwardly—the American model, some visiting academics, particularly Americans, have criticized the Japanese universities because they do not do what the better universities in the United States are doing. Such comparisons are useful, I believe, but imitation, once again, is an impossible and unrealistic goal. Just as a corporation in the United

States cannot be managed in a purely Japanese style, neither can a Japanese university educate its students with the curriculum and classroom style of MIT, for example. In Japan the university, too, is already a well-established social institution, rooted in a cultural milieu which has developed continuously for centuries. An institution, like a human being, can improve only by making the most of its own special characteristics. Demanding from an institution a new style of behavior that has no roots in the culture and is alien to the people who work in it will lead to frustration. Therefore, in my own critique of university education in Japan I have first restrained my tendency to compare it with the system in American universities I have known. Then I have tried to observe where and how the Japanese university—at least the one where I teach—is successfully educating students, and I will try to offer a few suggestions for extending its success.

With the passage of years—it was 1952 when I first came to Japan—I have become convinced that the two experiences which best succeed in educating the typical university student are participation in a student club and the graduation research project. After expanding a little on why I judge these two to be so successful, I want to offer a few suggestions on how university education might be restructured somewhat to build on these strengths.

Earlier in my teaching career I found myself telling students, especially the incoming freshmen, that it was dangerous to get too involved in a student club or circle. "You won't be able to spend much time or energy on studies if you get caught up in one of those student clubs or circles!" Fortunately, few students heeded my advice. Students have a healthy, almost instinctive sense that it is good for them to belong to a student club or circle. And when I think back on the students I have known, and especially after I see how they have developed in society after graduation, I must recognize that the shaping they received through the club was an important formative influence—perhaps one of the most important.

On the Sophia University campus, student clubs and circles proliferate like mushrooms in a rain forest. There are over 300 clubs and circles: for photography, ballroom dancing, rugby, off-road bicycle touring, mandolin ensemble, volunteer service to the handicapped, Andean folk music, movie appreciation, English-language drama, and English-language debating, and more than 30 circles for tennis—a rich profusion of youthful energy and developing talent. Largely outside the direction of the professors, these young adults, already at the age when personal aptitudes and intellectual acumen are reaching full flower, are molded in a social atmosphere which encourages initiative, allows scope for creativity, and provides just the right amount of criticism and coaching. I am constantly amazed at the high goals the clubs set for themselves. The camaraderie of the club or circle encourages and rewards committed effort while at the same time helping beginners progress beyond their first imperfect attempts.

Years ago I was amazed to learn that, for example, the main tennis club, which, like the "varsity" in an American university, represents the university in intercollegiate matches, admits as full members first-year students who have never yet held a racquet. If only the student commits himself to practice seriously and to take full responsibility as a member of the club, he is provided with coaching and encouragement to progress. Students who could play no musical instrument when they entered sometimes perform in a first-class student orchestra before they graduate. An incoming freshman who can hardly speak a simple sentence correctly in English is able to debate in English a few years later. There is in these clubs a secret of successful education which we teachers in universities should learn. I have corrected my judgment of what I, at first, considered to be just "playing around."

In my view of education, a young adult can only educate himself; his own self-initiated activity is all-important. The role of the teacher is to give some guidance and motivation for such activity: to coach and encourage, to correct, and to challenge the student to extend his talents to the fullest. The young adult student should be given as much autonomy as possible to develop his personal talent, but he also must be called to stick to his choices, to commit himself to what he has undertaken. Also, for sound development, whether in tennis or in mathematics, he needs friendly and frequent coaching. But he should be helped to grow to the point where he has the power to evaluate his own efforts and impose the needed discipline on himself.

The club gives the student what Professor Chie Nakane, a cultural anthropologist, has called the *ba,* the place and "frame," for his life and activity. Nakane has developed in her two books *Human Relations in the*

Vertical Society and *The Dynamics of the Vertical Society*[2] the hypothesis that the primary social unit, or "molecule," for all Japanese is the small social group within which vertical (senior-junior) human relations are far more important than peer relations. Japanese, according to her, give their allegiance as well as their time and energy primarily to only one group; one cannot be equally a member of several groups. In the club the student finds his frame, where he has his place in this small "vertical society." Here he finds familiar vertical human relations among friends both older (*sempai*) and younger (*koohai*). Here he feels at home and can engage in creative activity in an environment of common interest and shared responsibility.

The other very successful educational activity in universities is the graduation research of fourth-year students. In the electrical engineering department where I teach—and the practice is quite similar in other universities—students, just before the beginning of their fourth year are shown the list of research themes or projects that each individual professor proposes for senior student research. Trying to satisfy the desires of the students as much as possible, we assign about six senior students to each teacher. During their entire senior year the students assigned to me are "members" of my laboratory. A young faculty assistant also has his desk in my laboratory. There are also some graduate students who have been with me for two years or more. This group naturally forms the type of small group which Professor Nakane says is so natural and congenial to Japanese. This small group has an ideal ba, the laboratory, the scene and center of the group's activity and social life. The group is bound together by friendship, by healthy older-younger "vertical" relationships, and by a senior leader, the teacher, who, in Nakane's model, symbolizes and maintains the unity and smooth working of the group.

In the atmosphere of the laboratory, I have observed, the students im-

---

[2] "Human relations in the vertical society" is my translation of the title of Chie Nakane's book *Tate-shakai no Ningen Kankei,* Kodansha, 1967. Although not a translation, Nakane's book *Japanese Society,* University of California Press, Berkeley, 1970 (also Penguin, New York, 1973), seems at a cursory examination to cover the same matter. "The dynamics of the vertical society" is my translation of the title of Nakane's book *Tate-shakai no Rikigaku,* Kodansha, Tokyo, 1968. As far as I know there is no English translation of this book.

mediately feel at home: relaxed, accepted as individuals, incorporated into a primary "vertical society." This kind of ba ensures the essential human conditions for personal initiative and creative activity. In Japan it is only the very rare "loner" who can even get interested in studies which are not integrated into such a small-group atmosphere of friendship and cooperative effort. The goal, the theme of an individual student's graduate research, as in any true research, is usually not sharply defined, and so the student is challenged to creative self-determination. Also, commitment to his autonomously selected project is demanded of him. The group, by its friendly, relaxed atmosphere, encourages him and sustains him through the long, difficult efforts at research. He is not alone. And yet he has personal responsibility for selecting his goals and the means by which he will attain them. Finally, at every step of the way, he is corrected (mostly by imitating his seniors and emulating his peers), frequently coached, and his work evaluated.

Although the professor oversees the whole process, much of the education goes on without any need for him to be directly involved. The professor is free to plan the overall direction of the research and to advise at crucial junctures. In this atmosphere students who, in their classwork, seemed mediocre and colorless blossom into energetic and creative researchers. I have had students who significantly extended the research I had done in graduate school. One student, who among us teachers had the reputation of being a poor achiever (it was already his sixth year in the university), found and corrected a mistake in a long calculation I had thought to be correct. I had not asked him to check the calculation: he had done it on his own initiative just to understand better what he was doing in his own project. Some of the final undergraduate research reports I have seen are of a quality fit for publication as research papers.

Almost every year I am amazed at the rapid progress and great ability shown in this graduation research. And yet, I am also almost equally amazed at the passiveness and lack of interest in the lecture courses and the poor performance on examinations we professors give at the end of each semester. Are these the same students? What does the success of this graduation research tell us about how we might improve the rest of the process?

One of the greatest defects in the Japanese university is, I believe, that the students are "talked at" too much by professors. They have too little time and too little guidance and encouragement for their own personal study and practice. Too many hours sitting passively in the classroom starts a vicious circle, for the student has too little time and energy left for personal study. A student in engineering spends an average of six hours a day in the classroom or laboratory Monday through Friday, and possibly two to four more hours on Saturday. Also, students at Sophia University in central Tokyo commute an average of two hours a day on crowded trains and buses. Unless a student trains himself to "four-hour sleep" (the title of a book written by one of my colleagues), he cannot possibly get in three to four hours of solid private study each day. Many engineering students take almost no time for private study on an ordinary class day.

The result is that the student cannot personally assimilate what he hears in class. At the most, he carefully notes down what the professor says in order to review it the night or so before the examination. The teachers, on the other hand, realizing that the students have little time for personal study, tend to explain every detail in the classroom. The pace drops. Also the professor finds it almost impossible to promote individual activity by assigning projects and reports, because he has too little time and help to go over the papers and give feedback to the students. In Japan, where younger people learn so much from their immediate seniors in the small group, what a waste it is not to have the older students correcting the reports and homework of the younger students! But the greatest problem with the system of lecture courses is that the student belongs to no group, has no ba centered on curriculum-related study.

Thanks to the entrance examination system (called "examination hell" by the young), the typical engineering student enters the university with very finely honed skills in mathematics and exact knowledge of what might be called "factual" physics and chemistry. However, he also brings along a habit of studying only to master facts and acquire computational and problem-solving skills. From the very beginning the university should break that pattern to guide the student toward a more self-directed and critical method of study. In order to do this the strengths of the small social group and the vertical society should be brought into play.

Therefore, I propose that, immediately after entering the university, each student be incorporated into a small group (call it a "seminar-club") associated with one of the professors and the older graduate students under his direction. In my own department, for example, such a group could consist of about 24 students, six from each of the four undergraduate years, together with about six graduate students. The group would hold some seminar-style meetings, and each student would have his turn to make a presentation and have his work reviewed. The seminar presentation of undergraduate students would, for the most part, concern projects (homework, if you will) set by the professors giving the general lectures. In this way, the work of the student would be examined and corrected in the seminar group even before being given in final form to the professor in charge of a particular course. (I can hear an American reading this say, How will you evaluate the personal effort and comprehension of individual students if they get help from others? My answer is that grades are just not that important in Japan. There are other ways of evaluating each student—for example, by his contribution to the group.) The young student could also begin in some way to participate in the research of the group, especially during the long spring and summer vacations, which are otherwise lost—for academic purposes—during the first three years of the university.

This "seminar-club" would make it possible for all of the students to help in one another's education. The professor, the leader of the seminar-club, with his experience and larger view of the field, could contribute his best by not being overburdened with the many details that go along with direct supervision of each student's independent activity.

Then too the number of hours of lectures could be reduced. The lecturers could go ahead at a faster pace, satisfied that the students could master the material in private study and with the help of their seminar-club. The teachers would no longer have to presume (as they do now) that when the student is not in the classroom or laboratory, he is very likely with his club planning and preparing to cross the Sahara Desert on off-road bicycles next summer.

8

Family Considerations

Deborah A. Coleman Hann and Stephen A. Hann

When reviewing the original draft of this chapter, we found that it was essentially a 2200 word condemnation of high prices in Japan. We have decided to reduce our original tirade to a single section and devote the rest of the chapter to describing what our life has been like in Japan. We will restrict this discussion to what we have personally seen while living at our own expense in the Tokyo area for almost two years. Families with generous large-company sponsorship and the attendant benefits would have an entirely different story to tell.

JAPANESE PRICES

There is no point in repeating the familiar horror stories—although they are in large part true—about the astronomically high price of almost everything in Tokyo. A fairly reliable estimate for the cost of living in the Tokyo area can be arrived at as follows:

1. Domestic goods and services cost about three times what they would in the United States.
2. Imported goods cost about five times what they would in the United States.
3. The fees and deposits required to lease an apartment total a minimum of five months' rent.

Using these estimates when planning for living within commuting distance of Tokyo will probably provide a reasonably accurate budget for a foreign engineer and an accompanying family. Of course, many goods and services are reasonably priced in Japan, but the ones that are not tend to make the average costs adhere to these estimates. However, these estimates are based on a few assumptions:

1. Minimal purchases are made at trendy boutiques and top name stores.
2. Most entertainment at top clubs and restaurants is done at company expense.
3. One's apartment is at least a half-hour train ride from the city's business district.
4. The international schools are excellent but are affordable only by company-sponsored families.

There is simply no upper limit on how much money you can spend in Tokyo; salaried Japanese and self-financed foreign workers have no choice and abide by these restrictions. While the prices still cause us great discomfort, by applying discipline to our spending, we can afford to live in the greater Tokyo metropolitan area.

Having brought this issue out into the open, we will not dwell on prices in the rest of this article, but they are always lurking in the background as a trap for the unwary.

HOUSING

Upon arriving in Tokyo, we originally rented an 84-square-meter apartment in Tamagawa Gakuen, about 30 kilometers west of central Tokyo. It had three bedrooms, a living room, a dining room, kitchen, and one bathroom. It was on the first floor of a two-year-old six-story typical modern Japanese apartment building. It had carpeted bedroom floors, but no automatic dishwasher, and since we were the first tenants, we had to buy and install heater–air conditioning units (in Japan, central heating and air conditioning is not commonly used). The dining room was in traditional Japanese style with *tatami* (woven straw mats) as floor covers and *shoji* (sliding paper doors) to separate it from the living room. It was a 12-minute walk to the train station and a 45-minute express train ride to the nearest of the really important thoroughfares of Tokyo, Shinjuku. We found the neighborhood extremely pleasant and very much enjoyed the apartment, even though we picked it mainly because of the rent of 160,000 yen per month. The problem was that our complex was built into a hillside and the walk to the station meant climbing exactly 100 steps and then descending a steeply graded street. In temperate weather, which exists most of the time here, this was a minor inconvenience. But during the two or three months of hot, humid weather, we would arrive at the station soaking wet from the exertion of the walk. Also, it was a $1\frac{1}{2}$ hour trip, door to door, to many of the places in Tokyo where we normally go. Because of the hill and the wasted commuting time, we eventually decided to move someplace where we would have a flat walk to a more "strategically" located station.

After 16 months in Tamagawa Gakuen, we moved to a 100-square-meter two-story detached house in the town of Koganei, about 25 kilometers west of central Tokyo and about 10 kilometers north of our previous apartment. It is now a 6-minute, level walk to the station and a 30-minute local train ride to Shinjuku. The house also has three bedrooms (the middle-sized one has tatami and shoji, and the other two have bare wood floors and regular doors), a living room, a dining room (western-style), a kitchen, a carport, and $1\frac{1}{2}$ baths. The kitchen has ample storage space, but the range has no oven and only three gas burners. It costs 260,000 yen per month for rent, and we now have only two major housing problems. First, we need to buy rugs for the beautiful wood floors, and a dining room set. Second, we are very cold in the winter, because the stairwell, bathrooms, and hallway are unheated. It is standard in Japan to heat and air-condition only the main rooms, and only when they are in use. Other than that we are quite content with the new place.

TRANSPORTATION

The Japanese train and subway system is outstanding, since it is extremely safe and efficient. Trains run strictly on schedule, and they deliver you close to just about anywhere you want to go. Many foreigners, including us, have no desire to own a car and are perfectly happy with a combination of public transportation, walking, and taxis. In fact, we do not miss driving and are considering repatriating, when we return to the United States, to an area where we will rely less on cars. Even so, traveling from one place to another has been the largest adjustment that we have had to make in Japan. Part of the reason that we were initially overwhelmed by traveling in Japan is that Tokyo is the first really large city that we have ever commuted to. Having oriented our lives around cars in small and medium-sized cities in the United States, we never dreamed how much time spent climbing stairs and walking would be required in a mass transit–oriented society. Because almost everything we do centers on central Tokyo, we still spend far too much time on the trains. We expected central Tokyo to be crowded, but our expectations have been exceeded. The intense crowding on the major train lines during peak times makes it a chore to ride the train and walk near business-district train stations. But we never dreamed that the small-town train stations outside Tokyo would be so crowded. There are so many taxis, cars, bicycles, motorcycles, motor scooters, and pedestrians on the narrow streets around our new train station at peak hours that it is a safety hazard. To further complicate the traffic situation, there are no sidewalks along these streets. We doubt that people can be more aware of the crowding of Tokyo than when traveling.

SHOPPING

All of the necessities of life are available from the less expensive, second-tier stores. There is a domestic or licensed equivalent of virtually every western product, and so there are very few imported items available at most Japanese stores. And because of the small number of foreigners living in Japan, most imports are mainly suited to or modified for Japanese needs and tastes. This means that many foreigners, especially westerners, have trouble finding properly fitting clothes, their favorite foods, and many familiar housewares. Whenever we go overseas we bring back the maximum quantity of items that we miss from home but cannot find in Japan. In fact, on accompanied baggage, a surprising number of items carry little or no duty. However, as in all countries, the duties are inconsistent and it is best to look into the regulations before blindly bringing in large quantities of anything. Part of our two-year supply of asthma medication was refused entry as "excessive quantities." Some of our friends brought in a case of champagne and close to a case of good liquor and paid about one U.S. dollar per bottle in duties.

LANGUAGE, PEOPLE, AND CULTURE ⸺⸺⸺⸺⸺⸺⸺⸺⸺

Obviously, it is best to learn Japanese before moving to Japan. Predictably, very few foreigners have this skill when they move to Japan. Most Japanese have studied upward of half a dozen years of English from primary through secondary school. This is a misleading statement, because most of this effort is spent preparing for the national college entrance examination, which is a written test. Thus, relatively few Japanese are skillful at English conversation, even though they have some knowledge of English. Most engineers have a grasp of English, and all of our business is conducted in English. The second-tier stores where we do most of our shopping generally have no employees who speak English very well, and we end up using gestures, a few English expressions, and our very, very minimal Japanese.

The Japanese are hesitant yet curious when meeting a foreigner. There are very few westerners living in and around Tokyo and fewer still in the outlying areas of Japan. Except for those involved in international trade, very few Japanese are at all comfortable with foreigners. When dealing on a personal basis, they are polite and formal to the point where it gets in the way of trying to get acquainted. But in a crowd, especially on the platforms at train stations, all manners are suspended. The jostling can be unsettling, and no one hesitates to smoke in these crowds, even during the no smoking hours. But we have had Japanese strangers, both young and old, give us small presents in stores. There is a certain irony in that they do not know what to make of us and after two years we are still equally uncertain of them.

For our first six months in the old apartment, none of our neighbors paid any attention to us. After that, people in the neighborhood were greeting us and occasionally practicing their English on us. The constant invitations made so that people could really practice their English become somewhat of a minor irritation after two years or so of living in Japan. But the shopkeepers began to greet us, and we started to feel that we belonged there.

Since we moved to the new house, the people seem to be warming up to us a little more quickly.

CONCLUSIONS

It is our recommendation—and conversations with other foreigners lend further verification—that Japan should be regarded as an excellent place to pursue career and business interests. Additionally, it is possible, but takes considerable effort, to live a reasonably comfortable family life reasonably close to the western standard of living. Also, do not move to Japan without a preliminary house hunting and reconnaissance trip. We would have shipped more of our belongings had we known what we know now. Living in Japan is not for everyone, but with the right attitude and sufficient motivation it can be done. We expect to be here another two or three years, for a total of about five years.

PART

4

Cultural
Gaps and the
Language
Barrier

9

General Notes on Adapting to Life in Japan

Daniel K. Day

Living in a foreign country strains one's mental capacities in unexpected ways, and so I would urge anyone moving to Japan for a long period to read up on the country as much as possible before coming. I list some recommended readings at the end of this chapter; history and sociology are not everyone's cup of tea, nor are they mine, but knowledge of them in this instance is indispensable.

CULTURAL DIFFERENCES

I think that the most helpful attributes for non-Japanese wanting to live in Japan are intuition and a sense of humor. I will concentrate in the following paragraphs on negative things, since the unpleasant surprises are what newcomers need most to be warned about. I hope the reader can believe that there will also be many pleasant surprises. I love Japan and feel as much at home there now as I can imagine feeling anywhere.

Intuition

Intuition will be invaluable for telling you what is going on when the words are not enough; this is important, because the words can be deceptive if taken literally in Japan. As you will hear again and again, Japanese are concerned about maintaining harmony, or at least the appearance of harmony. This often results in people's not using the word for "no" even though that is what they mean. You must learn to read the signs of hesitancy and apparent waffling (and distinguish them from simple nervousness at dealing with a foreigner) which mean that a person is trying hard to think of an inoffensive way to refuse. A typical pattern of polite refusal found in the workplace is for someone to call over a senior, who in turn asks a coworker, who yet again asks his senior, who summons someone else, until you are surrounded by the makings of a committee whose purpose, you have dimly realized, is not to tell you "yes." This is neither waffling nor foot-dragging, it is protocol, and for you to bring it to an end before its time by blurting out an apology over your shoulder as you beat a hasty retreat would be unmannerly and counterproductive for the next time (you never know!) you go to the same office.

Haragei means the art (*gei*) of the belly (*hara*), where the belly signifies one's "heart," what one is really thinking. The art is in transmitting one's intention without putting it directly into words. The complementary process is to read (*yomu*) another's intention, and that is called *hara o yomu*. Humans everywhere do these things, but the Japanese language has an awe-inspiring number of expressions for them. This reflects, I think, the importance accorded to intuition in everyday dealings in Japanese society.

And sometimes Japanese people "lie." The first time you "smell a rat"— when someone listens to your rendition of a Japanese sentence, looks utterly mystified, and then remarks that your Japanese is really good—may mark your initiation into what Japanese call *honne* and *tatemae*. A common American phrase is "maybe that's what he said, . . ." with the pause indicating that whatever it was "he" said should be taken with a heavy dose of salt. This corresponds to the Japanese tatemae, while one's true intention (hara, in the previous paragraph) is honne. It would be unjust to get angry if you are a victim of tatemae; chalk it up to experience and keep trying to read the signs.

Slow Decisions

The breakneck pace of the urban centers notwithstanding, it is a given that group decisions in Japan take time. I have never worked in a Japanese company and cannot recall any anecdotes to illustrate this, but the western top-down management system which allows—indeed, forces—fast decisions does not exist here. Instead there is (have you heard this before?) decision by consensus, built painstakingly by those with the ideas by the process of persuading individuals to go along with their plan. The process is called *nemawashi*. Japanese are inured to these delays, even if they personally are irritated by them. You must adjust to them as well.

Reluctant Discussion

A westerner leading a group must adopt a different style when he or she wants people to contribute ideas. A common western pattern is for the leader to describe the plan of action and then to ask if there are any questions. This invitation will virtually always be greeted by a stony silence in Japan. This is partly because group discussions are not a part of the Japanese educational process. It is best not to expect members to speak out in groups at all; individually, though, people will feel more free to expose their ignorance . . . or brilliance.

Another facet of this, by the way, is one of the hardest things for non-Japanese to accept: the refusal of most people to discuss politics or anything else on which there could possibly be serious disagreement. This dovetails with the Japanese concern with maintaining harmony, as you will hear time and again; still, it may well be one of the weakest points of Japan as a supposedly democratic political unit, as all too many seem to refuse to take any kind of political stand at all. To the westerner who wants to discuss political and social trends and expects some educational disagreement as a matter of course, it can make the people, ultimately, a little boring to deal with on a personal level. Given linguistic skill and some time of acquaintance, though, some Japanese will open up.

LEARNING JAPANESE

There are a wealth of materials for learning the language; if you go to a school, your choice will already be made. If you plan to take lessons from a private teacher who is flexible about materials, I recommend the University of Hawaii Press series *Learn Japanese,* Volumes I through IV, for studying conversation. As for dictionaries, *romaji* (Roman alphabet) dictionaries featuring alphabetic renderings of Japanese words with their English equivalents will be valuable at first, but are usually rather limited in vocabulary. Kenkyusha and Sanseido each put out handy English-Japanese/Japanese-English volumes for about 2500 yen. The Japanese-English listings are given in *kana* (Japanese syllabic character) order; this will be bothersome at first, but will build your reading ability in kana and will drill you in the order, helpful once you develop the confidence to use directories, in buildings, for example. The English-Japanese listings will usually yield a line of kana and some frustrating *kanji* (Chinese characters), but sometimes you can simply show the printed page to the person to whom you are talking.

Do's and Don't's

I have taken few lessons in Japanese since coming to Japan, but judging from those and from my experience on the other side of the desk, I am confident about recommending some general do's and don't's. (1) You should set and keep in mind your goal in learning Japanese. How much do you want to learn, and how fast do you want to learn it? This is essential when making hard decisions whether a certain class or teacher is providing what you want and need. (A maxim: The object is communication in Japanese, not perfection in Japanese.) The ideal would be to observe a teacher at work before laying down any money, but I hear that most schools refuse to let prospective clients watch classes. Corollary to this is: (2) Do not allow schools to place you with students of a far different skill level. Your company may bring in a teacher and prefer to put everyone together to economize. Try to convince your manager that the company's money is best spent on lessons that are not a constant source of frustration to one student or another. (3) Small group lessons (two to five students) are the best; the individual gets plenty of attention, but has a chance to relax while other students interact with each other or the teacher. (4) Don't pay attention to the people who tell you "they all speak English." In the first place, they don't. In the second place, it is common courtesy to try to learn the language of your host country. In the third place, it will open doors, here in Japan and everywhere else there are Japanese, who have an enormous respect and gratitude for people who make the effort to learn their tongue. In this, they are no different from many peoples of the world, with the sad exception of the Americans, who have a wondrously arrogant lack of appreciation of the efforts of foreigners to learn English. (This, I think, is one of our mental blocks in studying languages; we fear that our work will go unappreciated.) (5) Don't believe people who tell you just to use the short forms of the verbs (you will understand what this means after you have studied for a few months) because "nobody uses the long forms." Oh, yes, they do. An explanation of these is outside the scope of this book; let it suffice to say that too much use of the short form is as irritating in Japanese as too much use of profanity is in English. (6) *Nihongo* is not the exclusive possession of the Japanese people. Whatever you learn is *yours*, so fine-tune your vocabulary for your own purposes. Make extra efforts. Look up words in your dictionary and learn vocabulary that is necessary and interesting to you. Try to express your sense of humor. (7) Sports are one of the very best ways of meeting Japanese in informal situations that do not involve drinking. Too often, conversations between Japanese and foreigners get onto one or the other of the languages themselves as the subject. Conversation, and your relationships with Japanese people, become more meaningful when you meet people in situations where you are *doing* something.

My Principles in Studying Japanese

Learning and retaining vocabulary is the final challenge of any language; you will be working on vocabulary long after you have attained a command of the grammar, and so I would advise a frontal assault on this. Different studying techniques work for different people, and so more important than my ideas is to be creative in trying to find ways that will work for you. Try not to overdo any particular method.

Vocabulary.   The primary rule is to learn first to *recognize* the meaning of a word when you hear it (write it down for study in *hiragana*, the kana for native Japanese words, not in kanji), and second to learn to *say* the Japanese word after seeing the corresponding word in your own language. Do not be ashamed or discouraged by your need to review constantly. My method is to make lists and carry them in a standard-size notebook or small memo book, glancing at them at odd moments and muttering them under my breath. Two of my friends who dislike lists have other methods. One carries around 20 or 30 scraps of paper in his pocket with the English on one side of a scrap and the Japanese on the other. He pulls out several scraps at a time and goes through them; when he has learned a word, he throws the scrap away. The other friend uses scraps of paper all together in a "grab bag" which he reaches into at odd moments.

The Kana.   After you learn how to write the hiragana, use them. Learn and write out the names of your coworkers, of stores and restaurants. There is plenty of geographical information which you will find useful in daily life, and it can be combined with practice of hiragana. Write out the names of subway stops, main streets, neighborhoods, wards, nearby towns, ski resorts, the 47 prefectures, and regions (Chubu and San'in *Chiho*, for example). Write out colors, tastes, shapes, textures, emotions. After you learn the *katakana*, or kana for western-derived words, copy down the entire contents of the menu one morning as you sit in the coffee shop. Take home a paper menu from Kentucky Fried Chicken and copy it. Write down all the animal names which should be written in katakana (*hamustaa*, for instance), the names of the elements, countries. This practice will also help you to remember how to pronounce these words, which are *Japanese* words and should be pronounced in the Japanese manner, as I note later on.

The Kanji.   The first goal in learning kanji is to be able to read signs. You need numbers and prices (15 kanji are used for these; they're an easy place to start); next, the days of the week (8 kanji). After these, you need addresses: learn to write your own address, then the other wards in your city, main subway and JR (Japan Railway) stations, and nearby towns. Note casually what individual kanji mean, but concentrate on your goal, addresses. Always try to write the strokes in the correct stroke order; this will help you eventually when puzzling out someone else's scribbled handwriting. Signs in the train stations will begin to make sense. The visual noise on business cards will resolve into real information. People's last names, which are generally based on place names, will begin to look familiar. If your interest in kanji is piqued by this, invest in a copy of Gakken's excellent kanji dictionary-textbook and ask your teacher or a friend to pick out a few vocabulary words under each character as useful for you to learn. As you do this, give priority to learning the word in its spoken form. This sounds contradictory, but as you practice learning to say the word while writing it, you will surely learn to read it. A word you can say and read but not write is worth more than a word you can read and write but not pronounce. Try to learn at a leisurely, steady pace, say, two kanji (six to eight vocabulary words) per day, five days a week. Review constantly. Kanji must be *written* (not just stared at) to be remembered, while vocabulary must be *vocalized*.

Pronunciation.   Again, the maxim: Your object is to learn to communicate, not to be perfect. That said, it is worth pointing out some pitfalls. Recent language-learning theory holds that there is a psychological "monitor" which operates while one is learning and speaking a foreign language and spots mistakes, or, on the other hand, refuses to spot mistakes. Adults tend to feel ridiculous imitating sounds which do not exist in their own language: the monitor malfunctions. This self-consciousness, if you suffer from it, is a delusion and should be fought with determination. Second, modern Japanese has many katakana words which were borrowed from English. These must be pronounced in the Japanese fashion, not the English. Imagine someone saying "hors d'oeuvres" in its French pronunciation in the midst of an otherwise English sentence, and you can probably understand how comical it sounds to mix pronunciations.

SUMMARY

There is no substitute for being able to speak Japanese. The writing system and the completely different grammar make for a formidable challenge, but they can be mastered with perseverance. I firmly believe that anyone who makes this effort will never regret it.

References

I recommend these books (and one article) on Japan, in order of priority:

Edwin O. Reischauer, *The Japanese,* Harvard University Press, Cambridge, Mass., 1977.

Boye De Mente, *Japanese Etiquette and Ethics in Business,* 5th rev. ed., National Textbook Company/Passport Books, Lincolnwood, Ill., 1986.

Frank Gibney, *Japan: The Fragile Superpower,* NAL Books, New York, 1989.

Karel van Wolferen, *The Enigma of Japanese Power,* Knopf, New York, 1989 (also Random House/Vintage, New York, 1989).

Akio Morita, *Made in Japan,* New American Library/Signet, New York, 1988.

Richard Mason and John Caiger, *A History of Japan,* Fress Press, New York, 1973.

Linda Sherman, "Breaking the Intimacy Barrier," *Japan Quarterly* (*Asahi Shimbun*), July–September 1990.

Textbooks for the Japanese language:

John Young and Kimiko Nakajima-Okano, *Learn Japanese: New College Text,* Vols. I–IV, University of Hawaii Press, Honolulu, 1984–85.

Kuratani et al., *A New Dictionary of Kanji Usage,* Gakken, Tokyo.

10

The Cultural Gap Experienced by a "Gaijin" Engineer in Japan

Raymond C. Vonderau

INTRODUCTION

In this presentation I will discuss my experiences and observations as a western-trained engineer in Japan that illustrate differences in practices in the engineering profession between Japanese and western society. The sum of these differences as they affect any one individual constitutes what is referred to in the title of this chapter as the "cultural gap."

The foreign-trained engineer can minimize the effect of the cultural gap by knowing about these differences in social culture (personal life) and corporate culture (working life). Through knowledge and understanding of each other's cultures, progress toward a *global* engineering community can be achieved. As the need for more foreign-trained engineers increases, it is hoped that this knowledge and understanding will help us to produce a group of professionals who can be considered *international* engineers rather than foreign engineers.

In identifying these differences, it is not implied that the Japanese way or the western way is wrong; however, these differences can cause difficulties for the foreign engineer which may keep him or her from performing as anticipated and from enjoying the opportunities provided by the new environment.

In discussing the cultural gap with both western-trained and Japanese engineers, I found that it had many different associations for different people. Since the experiences making up the cultural gap are very personal, each person is affected in a different way depending on his or her personality, interests, and particular situation. By receiving input from others, I hoped to present a broader range of experiences and opinions than only my own. Therefore I am indeed grateful to those people who shared their experiences and opinions with me in confidence.

My experience in Japan is unique, as is each foreign engineer's experience. I work alone in the offices of a major Japanese corporation which has been a licensee of my company in the United States since the 1950s. I am the ninth person from my company to be located in this assignment. Therefore my daily experiences are the result of a mature intercultural relationship between our companies. My experiences could thus be quite different from those of an engineer who is the first foreign person to work in his or her Japanese company.

Most Foreign Engineers' Assignments Are Temporary

There seem to be relatively few foreign engineers in Japan. The ASME Japan Chapter has about 350 members, of which only 20 members are foreign engineers, and the JSME has about 40,000 members, of which fewer than 100 members are foreign engineers. The assignments of these foreign engineers are somewhat varied, but a majority of them seem to be in academic and research fields in Japan. These assignments appear to be temporary in nature compared with a line assignment in the engineering department of a Japanese corporation. This is interesting because it seems to reflect the difficulty of making a permanent commitment to the "corporate fraternity" of a Japanese corporation, from which Japanese engineers have expectations of lifelong employment and regular advancement in job position along with their peers.

Hiring Practices

If one understands the Japanese educational system, the practice of hiring only graduating engineers into the company (who traditionally begin their employment on April 1 as freshmen) becomes quite clear; and if one understands the strong ties that develop between individuals and their company from this practice, then the difficulty of including foreign engineers in this system, as it exists, seems inevitable.

The larger Japanese corporations will also solicit and hire experienced engineers in order to have their expertise available for specific needs or projects. Joint ventures between Mazda and Ford and between Mitsubishi and Boeing are typical examples of such projects. Even though the scope of these projects may extend over a period of several years, this still falls short of a lifelong commitment to the corporation and for all practical purposes is a temporary assignment. I would conclude that, at this time and for many reasons, it is unlikely that many foreign engineers will make the lifelong commitment to a Japanese company that is necessary for inclusion in the Japanese "fraternity," with its benefit of job advancement.

Therefore, it seems that foreign engineers and Japanese engineers must learn to work effectively together even though the foreign engineers are not included in this fraternity. For a foreign engineer to be comfortable and effective in his or her job position and assignments, it is important that he or she understand the Japanese engineer's relationship to other members of the company and particularly how the Japanese system works. There is no written code of ethics that dictates the expected behavior from this relationship; such a code is not normally needed when practices stem from cultural roots. The foreign engineer, then, is obliged to learn through observation and perception about the pressures that mold corporate behavior.

THE JAPANESE COMPANY ENVIRONMENT

When observing Japanese engineers to learn the Japanese way of working life, the foreign engineer sees many activities, relationships, and peer pressures that differ from those experienced in western society. The most obvious difference, the first to be seen, is the arrangement of the office. In a Japanese office all the desks are arranged close together in clusters in a large room with relatively few or no separate enclosures. This makes working conditions very crowded, noisy, and disruptive of concentration. The large room appears to be difficult to heat uniformly in the winter, since it is usually too cold for comfort. Similarly, in the summer the room is usually too hot for comfort. Not every desk has a telephone, and to make or receive a phone call it is necessary for most of the engineers to leave their desks and use a phone at a centrally located desk. The foreign engineer can adjust to these differences, but until that happens, his or her contribution may be adversely affected.

Meetings

Many meetings are held during the working day to deal with any number of items of general company business: for example, weekly safety meetings, earthquake or fire drills, meetings for issuing routine announcements or planning company social projects, and periodic meetings to review the status of projects. The western engineer will soon note that there is very little participation by the individuals attending these meetings. This is especially surprising with regard to the project review meetings, since westerners are accustomed to using such meetings as an opportunity to contribute to the project. In Japan, however, the project group leader or manager does most of the talking during these meetings. Seldom is any criticism introduced, and seldom are alternative suggestions voiced which would have an impact on the project objectives or change the scope of study. This makes it difficult to know the true feelings of the Japanese engineers involved with the project.

Importance of Group Acceptance

To the foreign engineer, there appears to be little opportunity for individual creativity, since the total group has the responsibility for its members' contributions to their assignments and to project activities rather than the individuals themselves. Members of the group tend to become specialists in a relatively few disciplines. It also appears that members of the group are never aggressive but always conform to the expectations of the company, managers, and peers. It is obvious that acceptance and support from the group are very important to the Japanese engineer. Nothing is done to jeopardize this relationship. Therefore the working day is long—usually 10 to 12 hours, in order to strengthen the bonds holding the group together, even though not all that time is necessarily spent productively.

Vacations and Personal Time

Members of the group seldom take their allotted vacation time; one of the primary reasons seems to be the need to show support for the group's objectives. Japanese engineers are rarely away from the office except when they are traveling on company business or when the office is closed on Sundays and holidays. As a result of this peer pressure, the foreign engineer will notice that married Japanese engineers have little opportunity to spend time with their families, except perhaps on Sundays. These demands on a husband's time (as is usually the case) require strong support from his wife to run the household and to provide the necessary parental guidance for their children. In contrast, a foreign engineer is expected by his or her company to take any allotted vacation time.

PLANNING AND FOLLOWING A CAREER PATH

Several times in the past few years, I have asked young Japanese engineers what their career goals were and whether they had a plan or timetable to accomplish their goals. Nearly always the response has been that they have no personal goals or ambitions but rather only wish to be successful within their company by doing as well as possible in each of their assignments. Japanese engineers seem to be very willing to leave their career path planning totally in the hands of the company even though the company's plans are not made obvious to them. In contrast, the western-trained engineer is encouraged to set personal career goals and only to solicit the company's support to accomplish them.

The practice of annually promoting or moving people, particularly in large Japanese corporations, offers some opportunities for changing job assignments; however, many times changes are made without consulting the individuals affected by them. This can create problems for the individual and sometimes results in separation from the family when the new assignment is in another city or in a foreign country. It seems to be the case that employees do not refuse a company-planned career move, because doing so can seriously jeopardize opportunities to be considered for future assignment changes or promotion within the company.

On the positive side, the frequent movement of people within the company, coupled with lifelong employment, provides an excellent way to train company-oriented and knowledgeable managers. Strong managers who can lead people to accomplish the company's goals are one of the very positive results of the Japanese way of career planning, which allows the strongest person to achieve the position of leader of the group.

SOME DIFFERENCES THAT CAN CAUSE PROBLEMS

At the same time, there are some differences in customs that do continually cause difficulty for the foreign engineer in the Japanese work environment. One of these differences is that to be successful in a Japanese company, a foreign engineer must be able to work alone. This is partly because of the language barrier and partly because of the work ethic of the Japanese. There appears to be very little fraternizing or sharing of knowledge between peers even within the same group. Knowledge is shared more readily with subordinates, on an "as needed" basis, but with peers it is rarely shared even on an informal basis.

Demands on Employees' Time

In most Japanese companies, there are many traditional company-sponsored activities that are essentially social activities, since they are held outside the company. These are often ambitious programs, and the participation of each group member within the department is expected. Because the programs are ambitious, time for planning and preparation is required, which is normally taken out of busy working hours. These activities enable the employees to know each other better, and when the events are competitive, esprit de corps and a camraderie can be fostered; this is expected to enable the employees to work together in better harmony. By assigning different leaders or planners each year for these traditional events, some insight into each person's interest in and ability for leadership can be obtained. This can be helpful in identifying the "natural" leaders when the time for promotions arrives.

This total commitment of time—not only the lengthy workday, but also these additional company functions—is very difficult for the foreign engineer. Participation in these events, in addition to their working hours, makes it difficult for foreign engineers to find time to pursue any hobbies or activities of personal interest which are not related to the company or its activities. Therefore, foreign engineers may be concerned about becoming one-dimensional persons. The freedom of choice enjoyed in the home country is not available; one cannot pursue activities for personal growth and satisfaction, because the time is not available. The total commitment of time for the company's needs and objectives is difficult to accept for the foreign-trained engineer.

Differences Outside the Workplace

The foreign engineer can normally learn about and accept the Japanese customs in the workplace that have been described in this chapter. However, in order to fully enjoy living in a fascinating country with people of a different culture, it is also necessary to accept and adapt to many differences of a personal nature. As previously stated, the foreign engineer's stay in Japan is usually for a relatively short time. For this reason it may actually be easier to adjust to the kinds of differences in everyday living conditions that guidebooks often refer to as "culture shock." Some important concerns are as follows:

▶ The very crowded conditions and close proximity to people encountered in living, working, commuting, shopping, and enjoying recreation will cause mental fatigue in the foreigner first coming to Japan.
▶ The cost of rent, food, restaurant meals, and all Japanese goods and services is very high; however, it is also obvious that the quality of Japanese goods and services is very high.
▶ In most cities it is not possible to conveniently drive a car to accomplish personal errands, and the newcomer should accept the fact that ownership of a car is not always practical or necessary.

There are also many differences of a personal nature which make Japan very attractive to the foreign person:

▶ The politeness, courtesy, and work ethic of the Japanese people
▶ The many closely knit families with happy, well-behaved children
▶ The safety one feels in any part of the cities
▶ The efficiency of the transportation system

▶ The cleanliness of the Japanese people and their cities and living environment

▶ A strong respect for nature and a desire to preserve the beauty of the land, mountains, and sea

▶ A fascinating culture that teaches self-discipline and respect for others

In addition to the professional benefit of working in a Japanese company, all these are things a foreign engineer can enjoy by making the effort to bridge the cultural gap.

THE FUTURE: NARROWING THE CULTURAL GAP

If the shortage of technically trained professional people in Japan continues or becomes more acute, more foreign engineers are likely to be needed by Japanese companies. As suggested previously, it is likely that the foreigners hired will continue to be experienced engineers who are selected for having particular knowledge useful for a specific project need.

Considerable improvement in the necessary cultural adjustment is possible if these candidates can be taught—before they come to Japan and again after they arrive in Japan—about the known differences between Japanese and western society that will be encountered in living and working. Besides shortening the personal adjustment period, better knowledge of the cultural differences would enable the newcomer to make an almost immediate contribution to the group's project objectives.

Better knowledge also means an understanding of the cultural differences on the part of Japanese engineers and managers. Their interest could minimize the effect of these differences by leading management to make just enough changes in the traditional company system to facilitate the integration of the foreign engineer into the Japanese company family, even though he or she may have no intention of joining the Japanese "fraternity" of lifetime employees. One Japanese "sponsor" family could be appointed to assist and monitor each foreign family's acceptance and understanding of this new social culture; interested company members could volunteer. In addition, a company or corporate sponsor could be appointed to assist and monitor each foreign engineer's understanding of the corporate culture and the Japanese way.

The advantages to the Japanese company and to the foreign engineer in narrowing and bridging the cultural gap will be a more productive, more company-oriented, and totally involved *international* engineer. I think that this approach might provide a blueprint for a technical community that works together *on a global basis*—across national boundaries.

11

Be Wary of Generalizations, Be Humble, and Be Fearless of the Language

Craig Van Degrift

Understanding Japan is like examining an onion: one layer after another is probed only to reveal a deeper layer inside. The visitor passing through hastily on a business trip sees a western-style country that is peppered with shrines, temples, and castles which hint of a mysterious past. The Japanese seem to match their reputation for being a homogeneous swarm of tireless workers single-mindedly dedicated to their jobs. Even though their language is said to be the most difficult in the world, their literacy rate is the highest.

Visitors staying for a month will have explored a little deeper and gained some insights missed by the one-week visitor. If they stay for three months, they finally realize that understanding Japan is much more difficult than they had thought. After a year, their Japanese language study is likely to have made significant progress and they are likely to have made Japanese friends with whom they can have detailed discussions of political and cultural questions (probably still in English). If their friends do not represent too narrow a slice of Japanese society, they can finally start to really understand Japan.

The different writers' contributions in this book should help speed up this process, but the reader must be careful to remember that Japanese society is *not* perfectly homogeneous. Different observers' views will be shaped by those individuals with whom they interact as well as by their own individuality. *One must be wary that the unavoidable generalizations used to describe the traits of Japanese society as a whole not be blindly applied to individual people or a specific workplace.*

The views I present in this chapter are those of an American government scientist who worked for a year in Tsukuba at the Electrotechnical Laboratory (ETL) of the MITI (Ministry of International Trade and Industry), a particularly western-oriented laboratory in a newly created "science city." Prior to my arrival, I already had a number of close friends among Tsukuba's scientists, because my home laboratory, the National Bureau of Standards (now called the National Institute of Standards and Technology), regularly hosted Tsukuba scientists. My interest in the Japanese language had started 30 years earlier while I was in high school in Los Angeles, although it lay dormant during most of the interim.

I found the Electrotechnical Laboratory to be very similar to my home laboratory with respect to quality of equipment, facilities, and personnel. While individual offices were rare, everyone had his or her own telephone and adequate storage space. The group offices were relatively quiet and comfortable (our standards work required air conditioning), although there was a problem with "secondary" cigarette smoke.

The Japanese often speak of their nation's lack of creativity, but this was not evident at ETL. Even though its scientists and engineers have been processed by Japan's strict school system, the particular atmosphere of ETL allows a rekindling of their temporarily suppressed innate creativity. (The unusual nature of ETL may in part be due to the fact that a rather large percentage of its scientific staff has spent a year or more at research laboratories in Europe or America.)

Similarly, the reputed homogeneity of Japanese society does not exist at ETL. There seemed to be no less individuality among ETL scientists and their families than among their counterparts in America—some worked 8 hours a day, others 14; few wore suits; some fathers helped change diapers at home, others didn't; some repaired their own cars (one owned a 1964 Volkswagen beetle); and the scientists and engineers included some women together with many men. One difference, however, was that the ETL scientists seemed to read more technical literature (largely in English) and to attend fewer conferences than their contemporaries in America: there were many informal study groups that gathered to work their way through books (usually English-language books) in order to gain expertise in new fields. The most remarkable difference between ETL and a similar American organization, however, was in the laboratory's management. The characteristically Japanese bottom-up decision making was apparent, as was the informal bypassing of some formal bureaucratic barriers.

Surely, the work environment at ETL is exceptional for Japan, but that it exists at all at a major government laboratory underscores the danger of trusting too heavily in the usual generalizations about Japanese society. Japanese society and school traditions do *tend* to suppress individualism and creativity, but they by no means eliminate them.

While social interaction involving alcohol is common, I found that my absolute abstinence from alcohol did not create undue difficulty. Nonconformity of a foreigner, if it is not flaunted, is tolerated, and perhaps even admired.

It is important to be humble. It is very easy to be wrong anywhere, but when you are wrong in Japan you are unlikely to be clearly corrected. In fact, strong expression of opinion is likely to create serious barriers to further communication. Never underestimate your unassertive Japanese colleagues. Try to convey an uncertainty about the opinions you express. Using "I wonder if . . .," "Could it be that . . .," "Is it possible that . . .," and similar expressions can greatly soften one's speech. This tradition may be changing: one of my Japanese friends (observing my children) seriously asked, "How do American schools teach children to argue?" He was genuinely concerned that Japanese children should learn how to have more vigorous discussions.

It is easier to be *genuinely* humble in your ideas if you think more deeply about a given topic. For example, if you find yourself thinking, "The Japanese really need more big discount stores," remember that low prices and wide variety are not the only factors to consider. Walking to the corner store is something your kids can do unsupervised, is healthy exercise, doesn't use gas, and allows you to get by with a more compact refrigerator if you have a small apartment. Also, in the United States as well as Japan, the mom-and-pop stores employ a segment of society which might otherwise be unemployed or on welfare. Questions that seem to have obvious answers in American society can turn out to be more complex when fully analyzed. Japanese consideration of the long-term effect of procurement and sales decisions may lead to business practices which seem peculiar to foreigners.

Part of being humble is being willing to physically help in preparations for an activity as well as the cleaning up afterwards. Custodial help does not exist in schools and is minimal in workplaces. Always be on the alert for ways in which you should be helping in maintenance and other group activities, both at your workplace and in your neighborhood. The cleaning of trash pickup sites in residential areas and certain school maintenance tasks that the children cannot do are performed by residents and parents according to an established schedule.

Many Japanese scientists and engineers can understand English if one uses clearly pronounced, idiom-free English supported by clear diagrams. Nevertheless, a modest command of Japanese will allow you to better communicate with the people in your firm who are likely to be non-English-speakers—the machinists, production workers, and secretaries, for instance—as well as allow you to handle your personal life independently. If each interaction with lower-level workers requires the help of a Japanese colleague, your value to the firm will be diminished.

Even though full fluency in Japanese requires more than five years of intense study, every effort spent learning Japanese will yield benefits. It is quite a shock to well-educated visitors to realize that they are virtually illiterate! Simply attempting to study Japanese is an important social courtesy. It indicates an interest in the culture, an effort at self-sufficiency, and a recognition that communications with non-English-speaking Japanese are valued.

One can first concentrate on learning useful subsets of the full language. By intensively using flash cards, the 46-character *katakana* syllabary used for imported foreign words can be quickly learned. Since there are more than 30,000 imported words written in katakana, this provides immediate help in understanding many signs and package labels and even in discerning the subject of written text. Some skill and imagination, however, are required to correlate the Japanese and English versions of these words. Furthermore, they *must* be pronounced in the Japanese manner to be understood!

Next, one must learn the other syllabary (*hiragana*), start building a core vocabulary of genuine Japanese words, learn survival phrases and basic grammar, and begin learning the Chinese characters (called *kanji*). The hiragana requires the same degree of effort as katakana, but provides only slight immediate benefit until the core vocabulary is learned. Even though Japanese grammar has many levels of politeness and usage subtleties, learning enough for the purposes of basic communication is rather easy. Also, it is easy to learn to pronounce Japanese understandably. Be sure to get the vowel sounds right (there are only five), but ignore the tonal differences some books and courses try to teach.

It is the memorization of the genuine Japanese vocabulary and kanji that makes Japanese difficult. The words have no connection at all with those of western languages, and because they are built up from a small number of basic sounds, they are especially difficult to remember and to discern aurally. You should consult a small conversation textbook for a suitable vocabulary list rather than try to find Japanese equivalents to your favorite English words. Whereas children seem to be able to rather quickly learn to pick out words they know from rapidly spoken monotonic Japanese, this skill is only slowly learned by adults. Progress is greatly hastened if one dives in, not worrying about mistakes. Social interactions with non-English-speaking Japanese at singing bars, in athletic activities, or in craft clubs, for example, are very effective.

Kanji present the greatest linguistic barrier to foreigners. Fortunately, since few Japanese words will be misunderstood if written in hiragana instead of kanji, and since Japanese word processors have become widespread, *you can postpone or even skip learning to write kanji.* Just learn how to count brush strokes (to aid in using dictionaries), and concentrate on reading. Learning the meanings of the first 100 or so characters is easily accomplished using flash cards and a good pictorial imagination. Knowledge of these first few kanji can greatly help you in navigating Japan's transportation system. Learning the pronunciations, however, is not so easy, because each character will usually have two or three different pronunciations depending on associated characters or text. The most serious hurdle to learning kanji, however, is encountered when spanning the gap between the first 200 kanji and the 600 or 700 level. In this range, it seems that characters can be forgotten as quickly as new ones are learned. Also, there are many characters which are easily confused, defy pictorial interpretation, or have ill-defined meanings when studied individually. Fortunately, as your capability approaches the 1000 mark, the characters once again become *easier* to remember. This happens because kanji have many subtle meaning and pronunciation interrelationships which are not apparent until a large number of base characters and character components are known. Hang in there!

References

Perhaps the widest selection of Japanese language learning materials may be found at Bonjinsha Co., Ltd., in Tokyo or Osaka at the following addresses:

Tokyo: Second floor, Kojimachi New Yahiko Building, 6-2 Kojimachi, Chiyoda-ku, Tokyo 102. Above a Dairy Queen (opposite a Taiyo Kobe Mitsui Bank office), about half a kilometer east of the Yotsuya subway and train station on Shinjuku Street. Phone: 03-3239-8673. Fax: 03-3238-9125.

Osaka: Second Floor, Iwamoto Building, 7-1-29 Nishinakajima, Yodogawa-ku, Osaka 532. On the second floor of a building between the Shin-Osaka train station and the Chisan Hotel. Phone: 06-390-8461. Fax: 06-303-7049.

In the following listings the most recent known prices are given for the reader's convenience. Although these are subject to change, they will convey some notion of the "relative" cost.

An excellent, though expensive, set of pocket books written to help foreign visitors with Japanese culture, geography, and language is the illustrated book series published by the Japan Travel Bureau, Inc., volumes 1 through 13 (volumes 11 and 12 are in French). The first volume is *A Look into Japan,* Books Nippon, 1986 ($9.95, ISBN 4-53300-30-79).

The best technical Japanese textbook is *Basic Technical Japanese,* by E. E. Daub, R. B. Bird, and N. Inoue, published for Japan and Oceania by the University of Tokyo Press (7500 yen, ISBN 4-13-087051-3; 0-86008-467-1) and everywhere else by the University of Wisconsin Press, Madison, 1990 ($35, ISBN 0-299-12730-3).

The best kanji-to-English dictionary is the *Japanese Character Dictionary with Compound Lookup via Any Kanji,* by M. Spahn and W. Hadamitzky, published by Nichigai Associates, Tokyo (4950 yen, ISBN 4-8169-0828-5 C0581).

There are numerous textbooks for learning general Japanese and books to aid in kanji study. Of these, I recommend *Japanese for Today,* published by Gakken, Tokyo (3500 yen, ISBN 4-05-050154-6 C0081), and Kuratani et al., *A New Dictionary of Kanji Usage,* also published by Gakken (4800 yen, ISBN 4-05-051805-8 C3581).

The most complete Japanese-English and English-Japanese dictionaries are designed for use by Japanese and therefore assume a complete knowledge of kanji. The beginner must select from one of the less complete, "Romanized" dictionaries designed for foreigners.

12

Bridging Gaps by Cultural Simulation

Kazuo Takaiwa

CULTURAL GAPS

Culture Shock and Cultural Friction

When people come into contact with natives of a different culture, they may feel uneasy and sometimes embarrassed. We call the uneasiness *culture shock*; some of the victims of culture shock suffer great anxiety and have to return to their home countries. Simultaneously, the newcomers may create a similar shock in their counterparts. A crunch may come: cultural friction. Friction may give rise to a sense of incompatibility, dislike, and, at worst, animosity. Once distrust has reared its head, it is time-consuming and expensive to wipe it out.

P. R. Harris and R. T. Moran report in their book *Managing Cultural Differences*[1] that "adjustment problems of Americans abroad are severe, and adjustment failures are costly in terms of economics, efficiency of operations, intercultural relations, and personal satisfaction with duty abroad." Moran and Harris reported that an American company dispatched about 300 engineers and their families to Iran and had to return 33 percent to the United States. Each family that had to be replaced cost the company $210,000 (p. 164.).

---

[1] Philip R. Harris and Robert T. Moran, *Managing Cultural Differences*, 1st ed., 3rd printing. Copyright © 1983 by Gulf Publishing Company, Houston, Texas. Used with permission. All rights reserved.

"Some anthropologists, such as Kalervo Oberg, consider culture shock a 'malady, an occupational disease which may be experienced by people who are suddenly transported abroad'" (Harris and Moran, p. 92). But "culture shock is neither good nor bad, necessary nor unnecessary. It is a reality that many people face when in strange and unexpected situations." Some of these people, if it is important enough to them, will nevertheless choose to make the most of the experience of another culture in spite of the possible dysfunctional effects of culture shock (p. 93).

Culture and Civilization

Ryotaro Shiba, one of the most famous novelists in Japan, discussed the differences between culture and civilization in an article that I would like to paraphrase here.[2] Civilization is rational, he said, but culture is nonrational. For instance, a red traffic signal tells all vehicles to stop and a green signal tells them to go ahead. This convention is used throughout the world. In the sense that traffic signals are rational and universal, they belong to civilization.

On the other hand, culture comprises conventions that are not utilitarian in purpose and are observed only by specific groups—for example, specific nations—and is therefore nonrational. One group's culture is not applicable to other groups. For instance, in Japan, it is deemed good manners when opening a *fusuma*, a Japanese paper door, especially when greeting guests, for the lady of the house to drop to her knees, place her right-hand fingers on the flat metal pull, place her left-hand fingers on the lower edge of the fusuma, and slide it back with both hands. Of course, she could open a door with one hand while standing on her feet, but this happens not to be considered good manners. This is culture. Nonrational rules are themselves part of the joy of culture and part of what gives it its meaning.

I think that this is a very reasonable summary of the subject.

Struggles between Cultures

There is no such thing as superiority or inferiority among cultures; there are only differences. For example, when Japanese people want to call a

[2]Ryotaro Shiba, *Yomiuri Newspaper*, April 20, 1985.

friend over, they stretch out an arm with the palm down and move the fingers, while Europeans and Americans perform the gesture with their fingers up. One is not better than the other, just different. Whenever a nation becomes overly proud and proclaims the superiority—rather than simply the uniqueness—of its culture, it seals its fate to be overthrown, no matter how great it may have become. Culture springs from the history and life of a nation in the past and is not enhanced by chauvinism.

Unfortunately, it happens all too often that arguments and rifts over superficial differences occur between partners in intercultural efforts. To promote understanding, the differences between Japanese culture and American culture was the theme during the recent Structural Impediments Initiatives attended by the United States and Japan, which had been called by the American government to address the issue of trade imbalances. (One major difference is that democracy does not mean the same thing in Japan as in the United States.)

One and Forty

The ratio of the population of Japan (123 million) to the population of the world (5.2 billion) is approximately 1 out of 40. "One and forty" is my name for my concept of the proper attitude that Japanese should show during contacts with the rest of the world, as explained in my book *Abu Dhabi de Kaita Seiko suru Kaigai Bijinesu* ("success in business overseas: notes from Abu Dhabi," in Japanese). The book reports my key experiences in 15 years of assignments in Singapore, the Middle East, Africa, Indonesia, and South America, where I dealt with cultural friction between Japanese (including myself), the local people, and third nationalities. The projects involved several hundred or sometimes over 1000 Japanese engineers and supervisors, and several thousand local personnel. My philosophy is that since the Japanese constitute only one-fortieth of the world's population, it is essential for us to adapt to the ways of the west once we have left our own country. Conversely, when in Japan to do business, it is reasonable for westerners to be expected to adapt to Japanese ways in business.

Local people in many parts of the world have become accustomed to international ways, that is, western ways of doing business, during the long colonial period. On the contrary, Japanese until recently had few experiences of trade with westerners because of their national isolation for 300 years. Japanese do not live in a society of contracts and job descriptions, but in a society of oaths. They are expected to behave on their own as good citizens. The small-group activities TQC and TPM (see Chapter 6) are not compulsory but are undertaken autonomously to promote productivity. Japanese social rules are quite different from western counterparts. That is part of the concept of "one and forty."

Island Tribes and Continental Tribes

The Japanese are an "island tribe" (to paraphrase the Japanese term) that has developed its culture in relative isolation from other nations for more than 2000 years. Though peaceful intercourse with China and Korea has been recorded since the beginning of the Christian era, foreign commerce was strictly controlled by the imperial government once it had established itself in the sixth century. Japan has experienced only a few invasions by other nations. The first known in history were the Mongolian invasions of 1274 and 1281. On November 19, 1274, the invading Mongolian forces were devastated by a typhoon, which sank many of their ships as they began their retreat by sea. The next attempt, in 1281, suffered a similar fate, losing over half the force of 100,000 under the swords of the defending Japanese troops and a huge typhoon. This typhoon was called *kamikaze*, or "the wind of the gods," by Japanese historians. (From this experience, the word *kamikaze* became a metaphor for "luck" in Japanese. It was that desire for luck that led to its use in World War II—called the "Pacific War" in Japan—which ended when two atomic bombs were dropped on the country.)

Invasions of other countries by Japan have also taken place at only a few times in history. History tells us that the first was when Hideyoshi Toyotomi, the chief minister of state, sent troops to southern Korea between 1592 and 1598. The second was an invasion of Korea during the Sino-Japanese War, 1894–95. The third foreign campaign was in the Russo-Japanese War, 1904–05; the fourth, the attack on Qingtao (Tsingtao) in 1914. The fifth was the invasion of Manchuria and the Chinese mainland in 1931–40. The last was during the Pacific War from 1941 to 1945, at the end of which 4.5 million members of the Japanese armed forces returned to their burned-out homeland from the overseas dominions they had held.

Most civilizations other than Japan's have developed on continental mainlands, where borders provided by nature are less definitive. The peoples of such nations are called "continental tribes" in Japanese; these civilizations traditionally depended on territory to feed their populations and sought to increase their territory when their populations rose. (Starting with the Edo period after since 1603, food shortages and population increases did not tend to be a cause of war within Japan. The food supply came under control of the landlords, who encouraged development of farmland to prepare against famine.) Thus, continental tribes were constantly preparing to wage war, or to protect themselves against invasion. There was antagonism of one people against another; nations could not do business with each other without negotiations and contracts.

Wa (Harmony) and Antagonism

The relations typical of "continental tribes" were not an important element in Japanese domestic history. The Japanese, as a relatively homogeneous population, were not in a position to invade neighboring tribes or to be invaded by foreign tribes. Thus, Japanese had to live and stay in the islands in harmony with each other. To break harmony was tribal suicide.

The maintenance of harmony in social life, called *wa*, was required by the first article of the constitution of Shotoku Taishi (574–622), the most famous prince and one of the greatest men of wisdom in Japanese history. The concern with wa is echoed in the Japanese proverb "silence is golden but eloquence is only silver."

Japanese sought to be a single culture and to use a single language in their territory to keep harmony. If members of a different cultural group arrived on the scene, the Japanese tried to assimilate them so as to erase the sense of incompatibility. They lionized the newcomers at first, but then turned on them and forced them to acculturate. The histories of Europe, Asia, and Africa, on the other hand, are histories of antagonism, opposition, and self-assertion, as described by Paul Kennedy in his famous book *The Rise and Fall of the Great Powers*.[3]

CULTURAL SIMULATION

Principles and Procedures

I have developed a computer program of cultural simulation for testing and training in the ability to adapt to a foreign culture. (I have applied for a patent from both the Japanese and American patent bureaus.)

On the screen there are still pictures and moving pictures of scenes you might see in the country of interest that might seem odd to an outsider. The pictures are followed by five sentences that make up a multiple-choice question about the scene. One choice is the reaction a native would have; it is the "right" answer. The others are graded by an expert system to show the test subject's "adaptability to the different culture (ATDC)." Thirty scenes make up one test, and their content depends on the target country, the occupation of the test subject, and the subject's past experience.

Depending on the subject's answers, the ATDC grade is at one of four levels:

1. Is ready to adapt to the different culture.
2. Has adaptability to the different culture, but needs some training.
3. Has adaptability to the different culture, but needs a lot of training.
4. Is not ready to adapt to the different culture.

In the cases that follow, the target country is Japan, the occupation is engineer, and the grade of the subject's experience is "elementary," meaning that the subject has never been in Japan or is in Japan for the first time now. Three examples are shown.

Example 1: CS-4, NOODLE. (See Figures 12.1 and 12.2). Table manners differ extensively among peoples. Westerners use knives and forks;

[3]Paul Kennedy, *The Rise and Fall of the Great Powers: Economic Change and Military Conflict from 1500 to 2000*, Random House/Vintage, New York, 1989.

Figure 12.1 Scene: eating noodles.

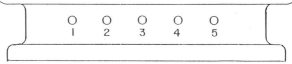

Figure 12.2 Answers for noodles scene.

Japanese, chopsticks; and Indians, fingers. The custom for drinking coffee and especially for taking soup varies substantially; the Japanese technique might seem particularly strange to a westerner and thus is a good basis for a test question. In the case of eating noodles, a respondent would be rated as follows:

▶ Answer 1: SWALLOW NOODLES WITHOUT CHEWING. If the respondent thinks that swallowing noodles without chewing is good manners, he or she has little understanding of Japanese manners, but may be able to achieve a high ATDC after much training. Therefore, the respondent is given a grade of 3.

▶ Answer 2: TAKING UP A BOWL AND PINCHING NOODLES IS BAD MANNERS. A bowl of noodles is heavy. In Japan, taking up the bowl would not be good manners. That would be more in keeping with Korean customs. The respondent will have good adaptability after moderate training and is graded 2.

▶ Answer 3: DON'T SIP SOUP NOISILY. Eating without noise is deemed most polite by westerners, but Japanese don't value it. The respondent needs a lot of training and is graded 3.

▶ Answer 4: EAT NOODLES OR SIP SOUP WITH A SLURPING NOISE. Japanese like to make noise while drinking tea, coffee, or soup. In the tea ceremony, specifically, the last sip should be accompanied with a sipping noise. Therefore, the respondent is a person who understands Japanese culture and is graded 1.

▶ Answer 5: NONE OF THE ABOVE IN JAPAN. The respondent has not found an appropriate answer among the alternatives. Thus he or she should have a very low ATDC and is graded 4.

Example 2: CS-11, PORCH.   (See Figures 12.3 and 12.4.) We invited a foreign couple to visit us. They came to our house and entered through the porch. I told them "Please come in." The lady then stepped inside without taking her shoes off. Usually, in Japan people go barefoot in the house. One of the greatest sources of embarrassment is that a foreigner is used to going into a house with shoes on. According to the program, a respondent will be rated as follows:

Figure 12.3 Scene: at the porch.

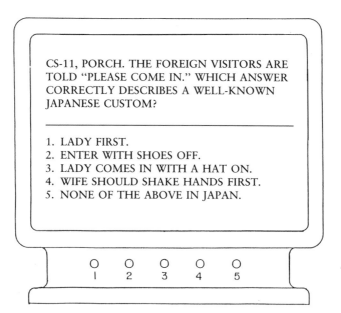

CS-11, PORCH. THE FOREIGN VISITORS ARE
TOLD "PLEASE COME IN." WHICH ANSWER
CORRECTLY DESCRIBES A WELL-KNOWN
JAPANESE CUSTOM?

1. LADY FIRST.
2. ENTER WITH SHOES OFF.
3. LADY COMES IN WITH A HAT ON.
4. WIFE SHOULD SHAKE HANDS FIRST.
5. NONE OF THE ABOVE IN JAPAN.

○ ○ ○ ○ ○
1 2 3 4 5

Figure 12.4 Answers for porch scene.

▶ Answer 1: LADY FIRST. The respondent receives an ATDC grade of 2.

▶ Answer 2: ENTER WITH SHOES OFF. This answer accurately reflects Japanese custom and will be given a grade of 1.

▶ Answer 3: LADY COMES IN WITH HER HAT ON. The respondent is graded 3.

▶ Answer 4: WIFE SHOULD SHAKE HANDS FIRST. The respondent is graded 3.

▶ Answer 5: NONE OF THE ABOVE IN JAPAN. This answer is given the lowest grade, 4.

Example 3: CS-19, SLIPPERS.   (See Figures 12.5 and 12.6.) In a Japanese house, people usually walk barefoot or wear either socks or *tabi*, Japanese-style short socks. Slippers are offered to guests to keep their socks clean. This causes another problem, because foreigners are not used to wearing such slippers in their home country. They usually wear shoes at home or when visiting. Furthermore, for the lavatory, Japanese provide a separate pair of slippers to be worn exclusively there. Because the Japanese regard

Figure 12.5 Scene: SLIPPERS.

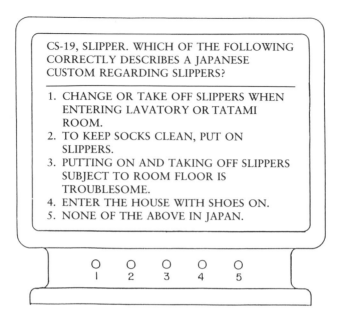

CS-19, SLIPPER. WHICH OF THE FOLLOWING
CORRECTLY DESCRIBES A JAPANESE
CUSTOM REGARDING SLIPPERS?

1. CHANGE OR TAKE OFF SLIPPERS WHEN
   ENTERING LAVATORY OR TATAMI
   ROOM.
2. TO KEEP SOCKS CLEAN, PUT ON
   SLIPPERS.
3. PUTTING ON AND TAKING OFF SLIPPERS
   SUBJECT TO ROOM FLOOR IS
   TROUBLESOME.
4. ENTER THE HOUSE WITH SHOES ON.
5. NONE OF THE ABOVE IN JAPAN.

Figure 12.6: Answers for SLIPPERS scene.

the toilet as a dirty place, these special slippers are never used outside the toilet room. Therefore, everybody is requested to change slippers upon entering the lavatory.

In a room where *tatami* (Japanese straw mats) are laid, the Japanese do not usually use slippers, because they don't like to see any dust that may be stuck to the underside of the slippers drop on the mats. It is troublesome to have to sweep the *tatami* later. Therefore, everybody is asked to remove his or her slippers when entering a room with *tatami*. This is a characteristic Japanese custom.

Using the grading system, the ATDC scores for Figures 12.5 and 12.6 are as follow:

▶ Answer 1: CHANGE OR TAKE OFF SLIPPERS WHEN ENTERING LAVATORY OR TATAMI ROOM. This is the Japanese custom. It receives an ATDC grade of 1.

▶ Answer 2: TO KEEP SOCKS CLEAN, PUT ON SLIPPERS. This partially describes the Japanese custom and is graded 2.

▶ Answer 3: PUTTING ON AND TAKING OFF SLIPPERS SUBJECT

TO ROOM FLOOR IS TROUBLESOME. The respondent shows great uncertainty and is graded 3.

▶ Answer 4: ENTER THE HOUSE WITH SHOES ON. The respondent is again graded 3.

▶ Answer 5: NONE OF THE ABOVE IN JAPAN. As before, respondent receives a grade of 4.

We prepared computer programs for this test and exhibited the system at a symposium in 1988. We are planning to use the expert system in the future to evaluate the ATDC.

Computer Analysis

We give a test comprising 30 scenes before a training program and another test after the program. The comparison of ATDC scores tells us individuals' rates of progress in cultural adaptation, while the average score difference tells us the effectiveness of the training. A more advanced course in cultural simulation can be taken afterward if desired.

By testing and improving their ATDC, foreign-born engineers can have an enjoyable stay and a productive working life in Japan.

The author wishes to express his gratitude to Daniel Day for his guidance in using the English language during the course of writing this article.

The Benefits of Professional Societies

13

Professional Societies Promote International Engineering Practice

Arthur E. Bergles

The 160 countries of the world are moving toward a global economy, with the prospect of one marketplace, in spite of the disruptive war in the Persian Gulf in 1991 and the serious internal conflicts in several countries. This movement is characterized by increasing sophistication of technology; indeed, it is technology that brings the world closer together. It is not surprising, then, that one of the most significant trends has been the globalization of engineering.

In the United States there are 125 national engineering societies representing most of the approximately 2 million engineers currently practicing in the country. The major societies have taken on a decidedly international character. For example, the Society of Automotive Engineers has changed its logo to SAE International, and its main annual event is the SAE International Congress and Exposition. The formal technical program for the February 1991 Congress and Exposition in Detroit indicates that 41 percent of the 1100 papers originated in organizations outside the United States, and 22 percent of them were from Japan. The American Society for Metals has become ASM International. The Institute of Electrical and Electronic Engineers has about 74,000 of its 320,000 members overseas. In turn, increasing numbers of American engineers hold memberships in foreign professional societies. The globalization of engineering has clearly brought with it the internationalization of engineering societies, and, indeed, these societies are leading the way toward the globalization of engineering.

The American Society of Mechanical Engineers provides a good example of this internationalization. ASME has evolved from its founding as a rather local organization in 1880 to a comprehensive technical society having worldwide activities and a truly global outlook. The total membership of ASME is 120,000, including 20,000 students, and there are about 6000 foreign members. The United States, Mexico, and parts of Canada are represented by 12 regions, 197 sections, and 295 student sections and clubs. ASME has concluded 25 agreements of cooperation with appropriate organizations in 21 countries, and has 7 operating overseas chapters and 26 correspondents.

Technical affairs are covered by 36 technical divisions and 3 institutes that cover the wide spectrum of specialties represented by modern mechanical engineering. For 1991, thirty-two technical conferences were scheduled, with four of them located outside the United States. The *ASME Transactions* now include 17 journals. In 1990, there were also 205 bound volumes of conference papers, 850 preprints, and a variety of books and other publications distributed worldwide. ASME is one of the world's major technical publishers. The 600 codes and standards include the comprehensive Boiler and Pressure Vessel Code initiated over 75 years ago and now used in 35 countries. Short courses that now number between 120 and 140 annually include overseas offerings.

ASME's relations with Japan have been particularly meaningful, with long-standing joint conferences, an active chapter (the ASME Japan Chapter), a corresondent (Mr. Haruo Kozono for the 1988–92 term), and ASME's support of the new *JSME International Journal.* Conferences in 1991 include the joint third ASME-JSME Thermal Engineering and ASME-JSME-JSES Solar Engineering Conference in Reno and the First ASME-JSME Fluids Engineering Conference in Portland, Oregon.

Because of these meaningful activities and resultant "networking" of engineers, it is possible for engineers taking on assignments in Japan to make professional and personal transitions more quickly. Chapter meetings provide the social contact, while newsletters—such as *ASME Worldwide* and ASME Japan Newsletter—and correspondent communications provide a steady stream of information. These local contacts are also effective in bringing the latest ASME technical information to the attention of members, an important consideration given the exponential growth of technical literature. Electronic media are becoming more important in transmitting technical information as well as information on society business. In general, ASME and other societies are actively developing programs that provide services both to overseas members and to American members working abroad.

Elsewhere in this book, the experiences of more senior engineers are documented. It is also of interest to mention the early career activities of an outstanding young engineer. Linda M. Strunk, an associate member of ASME, graduated from Rensselaer Polytechnic Institute in 1988 with a B.S. in mechanical engineering. Having an interest in an overseas assignment, she studied Japanese in Singapore during the summer of 1987 and at the State University of New York at Albany during 1987–88. Through a college job fair in Boston, she made contact with several Japanese firms, one of which sponsored an interview trip to Tokyo. In September 1988, she took a computer programming position in the financial systems department of Toppan Moore Systems, Ltd. Since September 1989, she has been a visiting researcher in the Hitachi Mechanical Engineering Research Laboratory in Tsuchiura. She is working in the field of intelligent robotics, and has collaborated on the design of a linearly mapping stereoscopic visual interface for teleoperated space manipulators. A paper based on this work was presented in July 1990 at the IEEE International Workshop on Intelligent Robots and Systems '90. She is now working on the force control of a dexterous robotic hand. Linda's job at Hitachi is a direct result of contacts made within the ASME Japan Chapter. She is a "global engineer" who has clearly succeeded in the practice of engineering in a foreign country. The extent of her accomplishment was demonstrated during a trip I made to Japan in 1990 during which my credibility was clearly enhanced once high Hitachi officials found out that I knew Linda.

It is evident from this discussion that professional societies such as ASME do promote international engineering practice. ASME has recently established a board on international affairs that coordinates agreements, chapters, and correspondents as well as international initiatives throughout the entire society. The fruitful cooperation with JSME and the numerous ASME activities in Japan are testimony to the success of this new global effort.

14

Welcome to Japanese Professional Societies: The Case of JSME

Hideo Ohashi and Hiroshi Honda

With the advancement of science and technology and industrial development, Japanese professional societies have grown steadily and extensively to meet the demands of Japanese. Among the numerous professional societies of various kinds in Japan, the engineering societies are considered to be among the most prosperous and active category, reflecting Japanese industrial success. Numerous Japanese companies support these societies professionally and financially as corporation members, while university professors play leading roles in their activities.

It can be seen from Table 14.1 that membership in major Japanese engineering societies ranged from 24,000 to 41,000 as of December 1989, with JSME having the greatest number (43,000 as of September 1990). However, the Institute of Electronics, Information and Communication Engineers has been the fastest-growing engineering society in recent years—which is not surprising, given the current impressive advances in electronics and communications technology worldwide and in Japan. All these societies welcome the participation of foreign-born engineers in their activities in a similar manner. In this chapter, the case of JSME is introduced as an example.

The number of members of JSME is about one-third that of ASME. Considering that the population of Japan is one-half that of the United States and that almost all JSME members are Japanese while ASME has a substantial number of members in Mexico, Canada, and other foreign countries, memberships of both societies as a fraction of national population are not far from the same level. To cope with the increasingly internationalized and globalized environments faced by our members and

TABLE 14.1   Major Japanese Engineering Societies and Numbers of Members (as of December 1989)

Japan Society of Mechanical Engineers (JSME)	41,000
Chemical Society of Japan	36,100
Institute of Electronics, Information and Communication Engineers	35,200
Japan Society of Civil Engineers	30,000
Architectural Institute of Japan	29,100
Institute of Electrical Engineers in Japan	24,000

their employers, JSME has concluded agreements of cooperation with ASME; the Institution of Mechanical Engineers (ImechE), U.K.; the Chinese Mechanical Engineering Society (CMES); the Verein Deutscher Ingenieure (VDI), Germany; the Korean Society of Mechanical Engineers (KSME); the Philippine Society of Mechanical Engineers (PSME); the Institution of Engineers, Australia (IEAust); and the Czechoslovak Society for Mechanics. Through cooperation with these societies, JSME has sponsored or co-sponsored international conferences such as those listed in Table 14.2 and will continue to do so. With the increasing number of foreign-born engineers living in and visiting Japan, JSME is welcoming the opportunity to exchange ideas and views and to build friendships with international engineers. To this end, it is a precious opportunity for JSME and ASME Japan Chapter to cosponsor forums such as The Role of Engineering Societies in the Age of Internationalization, which was held on March 31, 1990.

JSME also encourages "grass roots" regional activities at its seven branches all over Japan, shown in Figure 14.1.

Toward its centennial anniversary in 1997, JSME is building a new image while maintaining its traditional style. JSME published its first English issue of *JSME News* in 1990, offered various engineering courses that could lead to lifetime education, and adopted a slogan, "From power mechanics to intelligent mechanics," emphasizing the importance of the extensive application of intelligent technology to mechanical engineering. "Intelligence" has rapidly changed the content of mechanical engineering in recent years with the advances in robotics and mechatronics, for which JSME has an independent division. Divisions of ASME and JSME are compared

TABLE 14.2 Sample Meeting Calendar of JSME

1990	1991
May 29 to June 1. International Conference on Manufacturing Systems and Environments—Looking toward the 21st Century (Tokyo, Japan)	March 17–22 2nd ASME/JSME/JSES Solar Energy Conference (Reno, USA)
	April 21–25 International Conference on Computational Engineering Science (Patlas, Greece)
June 5–7. Japan/U.S.A. Boundary Elements Symposium (California, U.S.A.)	
July 6–7. 1st KSME/JSME Fracture and Strength Conference (Seoul, Korea)	June 18–20 Conference on Engineered Materials and Structures (Berkeley, USA)
August 5–9. 1990 ASME International Computers in Engineering Conference & Exhibition (Boston, U.S.A.)	June 23–27 ASME/JSME Fluid Engineering Conference (Portland, Oregon, USA)
	July 8–12 MECH '91—International Mechanical Engineering Congress (Sydney, Australia)
August 20–24. 9th International Conference on Experimental Mechanics (Copenhagen, Denmark)	August 29–31 3rd Triennial International Symposium on Fluid Control, Measurement and Visualization (San Francisco, USA)
August 26–31. 1st World Congress of Biomechanics (San Diego, U.S.A.)	
September 3–5. International Symposium on Diagnostics and Modeling of Combustion in Internal Combustion Engines (Kyoto, Japan)	September 2–13 ISO/TC108 Kobe Meeting (Kobe Japan)
	November 4–7 1st JSME/ASME Joint International Conference on Nuclear Engineering (Melbourne, Australia)
September 23–25. JSME Fall Annual Meeting (Sendai, Japan)	
October 11–12. KSME/JSME Fluid Engineering Joint Conference (Seoul, Korea)	November 23–29 The JSME International Conference on Motion and Power Transmission (Hiroshima, Japan)
November 13–15. 1st Joint U.S./Japan Conference on Adaptive Structures (Hawaii, U.S.A.)	November 25–29 Asia-Pacific Vibration Conference (Melbourne, Australia)

1991	1992
1991 January 28–31 International Symposium on the Application of Electro Magnetic Forces (Sendai, Japan)	April 9–12 ASME/JSME Joint Conference on Electrical and Electronic Packaging (San Jose, USA)
March 4–5 Australian Disaster & Emergency Management Conference (Sydney, Australia)	September 8–10 The 1st International Conference on Motion and Vibration Control (Yokohama, Japan)
March 17–22 1991 ASME/JSME Thermal Engineering Conference (Portland, Oregon, USA)	**1993**
	JSME/ASME Joint Power Generation Conference (Japan)

TABLE 14.3   Divisions of JSME and ASME

JSME	ASME
Materials and Mechanics	Applied Mechanics, Safety, Pressure Vessels and Piping, NDE Engineering
Dynamics and Control	Dynamic Systems and Control
	Noise Control and Acoustics
Fluids Engineering	Fluids Engineering, Fluid Power Systems and Technology
Thermal Engineering	Fuels and Combustion Technology
	Heat Transfer
Bioengineering	Bioengineering
Factory Automation	
Robotics and Mechatronics	
Computational Mechanics	Computers in Engineering
Engine Systems	Internal Combustion Engine
Environmental Engineering	Environmental Control, Solid Waste Processing
Power and Energy Systems	Nuclear Engineering; Power; Gas Turbine Institute; Advanced Energy Systems
	Solar Energy
Machine Design and Tribology	Tribology, Gear Research Institute
Materials and Processing	Materials
Manufacturing and Machine Tools	Production Engineering
Transportation and Logistics	Materials Handling Engineering
	Rail Transportation
Industrial and Chemical Machines	Process Industries, Textile Engineering
	Plant Engineering and Maintenance
Space Engineering	Aerospace
Design and Systems	Design Engineering
Technology and Society	Management, Technology and Society
	Petroleum, Ocean Engineering
	Offshore Mechanics and Arctic Engineering (OMAE), Mining and Excavation Research Institute
	Electric and Electronic Packaging
Information, Intelligence and Precision Equipment	

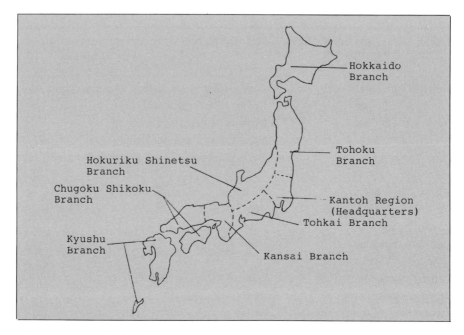

Figure 14.1 JSME's branches.

TABLE 14.4   JSME Meetings

Annual Meeting (April) in Tokyo
Fall Meeting (October) outside of Tokyo
Meetings, forums, symposia, lectures, organized by technical divisions
Meetings, forums, symposia, lectures, organized by regional branches

---

TABLE 14.5   JSME Publications

JSME International Journal, monthly, in English (submission is welcome)
JSME Journal, distributed monthly to members (list of contents in English)
JSME Transactions, monthly, in three series, A, B, and C (titles, authors, abstracts, legends of figures, and tables in English)
Conference papers (titles, authors, abstracts, figures, and tables in English)
JSME Standards (some bilingual)

S45C 調質材の腐食疲労過程
Corrosion Fatigue Process of a Heat-Treated 0.45% C Steel

正　後　藤　真　宏　（大分大）　　　松　田　裕　一　（多田野鉄工）
正　西　谷　弘　信　（九　大）　　　三　浦　篤　義　（大分大）

Masahiro GOTO, Atsuyoshi MIURA: Oita University, Oita, 870-11, Japan. Yuuichi MATSUDA:Tadano-Steel Ltd. Hironobu NISITANI: Kyushu University, Fukuoka, 812.

In order to clarify the corrosion fatigue process, rotating bending fatigue tests of heat-treated 0.45% C steel smooth specimens were carried out in 3% NaCl solution. The emphasis is to perform the successive observations by the plastic replica method. Results show that the corrosion pits are produced at the early stages of cycling, then most of them grow with further cycling until a crack initiates from each corrosion pit. The initiation of corrosion pits from slip bands is observed in only the case when the stress range is relatively large (in the range of stress under which slip bands are produced in air).

Key words: Corrosion Fatigue, Fatigue Process, Corrosion Pit, 3% NaCl Solution, Successive Observation, Heat-Treated Steel.

1. 緒　言

腐食環境下では、大気中の疲労限度よりもはるかに小さい応力のもとでき裂が発生・伝ぱし、部材を破断に導く。また、破断に至るまでの繰返し数も一般に大気中より小さい。したがって、腐食環境下の疲労被害を検討することは実用上重要である。本研究ではS45C調質平滑材の3%食塩水中における回転曲げ疲労試験を行い、レプリカ法による連続観察の結果に基づいて腐食環境下の疲労過程を明らかにする。

2. 材料・試験片および実験方法

素材は市販のS45C圧延丸棒であり、それを調質した後試験片を製作した。熱処理後の機械的性質を表1に示す。図1に試験片の形状および寸法を示す。また、一部の試験片には図2に示す形状の浅い部分切欠が加工してある。試験片製作後、電解研磨により表面を直径で約40μm程度取除いてから実験を行った。

使用した試験機は、小野式回転曲げ試験機である。腐食液は3%食塩水であり、腐食疲労の実験はこれを試験片中央部に流量約1000cc/minで滴下する循環方式で行った。液温は25±1℃に保ち、腐食液は各試験ごとに交換した。疲労被害の観察および腐食ピット寸法・き裂長さの測定はレプリカ法により行った。

3. 実験結果および考察

図3に平滑材のS-N曲線を示す。●印、▲印はそれぞれNaCl溶液中、大気中の部分切欠材の破断寿命を示す。

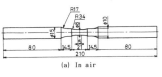

(a) In air

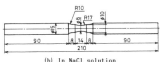

(b) In NaCl solution

Fig.1 Shape of the specimens

Table 1 Mechanical property

MPa			%	Hv*
σsl	σB	σT	ψ	
750	833	1510	61.6	270

*Vickers hardness (Load: 9.8N)

Fig.2 Shape of the notch

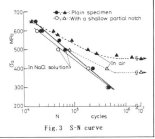

Fig.3 S-N curve

Figure 14.2 First page of a paper.

in Table 14.3, and it can be seen that the activities of both societies are extensive; however, the Electric and Electronic Packaging Division and divisions related to energy resources technology such as Petroleum, Ocean Engineering and OMAE are not covered by corresponding divisions of JSME.

JSME welcomes and encourages the participation of foreign-born engineers in its meetings. Foreign-born nonmembers are welcome to present their papers in English at the JSME domestic meetings listed in Table 14.4. In addition to *JSME International Journal,* published monthly in English, JSME makes the utmost effort to include English expressions in its domestic journals, transactions, conference papers, and standards, as shown in Table 14.5. Figure 14.2 shows an example of a conference paper that includes English expressions.

These efforts are also made by other Japanese professional societies in order to cope with the internationalization and globalization of the professional environment.

The authors wish to express their gratitude to Tsutomu Nakajima of JSME headquarters for his assistance during the course of writing this chapter.

PART

6

Employment
Case Studies

Perspectives of Industry

IMPRESSIONS FROM A ROCKWELL-HITACHI EXCHANGE PROGRAM

Peter E. D. Morgan

I had visited Japan four times, for periods of about two weeks each, before I went, in 1989, to work for six months at the Superconductivity Center of the Hitachi Corporation in Hitachi-*shi* (city), Ibaraki-*ken* (prefecture), located about 150 kilometers northeast of Tokyo on the Pacific coastline. My earlier visits had been primarily for scientific conferences, but each time I had spent about one week touring Japan and visiting colleagues, some of fairly long standing. I had taken a very short course in Japanese at Rockwell (spoken only); mainly out of curiosity about how the structure of the language differed from that of Indo-European languages. I found Japan to be extremely appealing in having attained success with a different mix of assumptions from those implicit in the western model and was greatly intrigued by the different weight given to ancient and modern ideas and traditions. To realize that religious concepts and customs, akin to those of the Greek and Roman civilizations—the western fountainheads—are alive and well today,[1] coexisting with influences from India, Korea, and China, . . . that's a long story.

At school in my native Great Britain, I had studied French and Latin

---

[1]Space prevents detail, but modern manifestations are, for example, the Shinto blessing and inauguration ceremonies for the installation of important new pieces of scientific equipment and for the sealing of significant business contracts.

and so was not nervous about attempting to learn another language; in fact, I regarded it as a challenge, anticipating the language to embody, as is usual, something of what makes its associated culture distinct. In any event, what I had learned from the trips and the pathetically little language I already knew encouraged me toward more effort; reading literature on Japan, I was particularly impressed by the work of Lafcadio Hearn (alias Yakumo Koizumi), a rather curious Britisher (of Irish-Greek extraction) who arrived in Japan in 1890, never again to leave, and who insightfully documented Japanese customs, mores, folktales, and ghost stories, among other things. He wrote in English, I believe, but now, translated into Japanese, his work is part of the lore of Japan. I felt an accord with his observations that reflected, I imagined, a common upbringing and experience with that other island people at the opposite end of the Eurasian landmass.

Space does not permit elaboration, but you will perceive very personal reasons for my affinity with Japan; I recommend that each individual contemplating ties with Japan, visit first, thus finding his or her own inclinations for wanting to forge a closer relationship. There was a cultural intention to my desire for more experience of Japan as well as a professional, engineering, and scientific significance. Cultural aspects are rather pervasive in Japan (as they, at least, used to be in Great Britain), and the Japanese hardly expect to associate only in a professional way, but expect also a more personal and intellectual transaction.

This prologue will suggest that when asked by my company, Rockwell International, if I would like to spend six months with Hitachi, I was predisposed to welcome the chance. When asked later how I had felt when the opportunity arose, I replied that it was as though an actor (I live near Hollywood) had been offered a part in a Kurosawa movie. Any good actor would give his eyeteeth for such an offer. Please don't misunderstand this point—I don't mean to say that I intended to play a role other than myself, but rather to see if I could fit into the very different cultural backdrop and different customs and even benefit the overall drama. In other words, I welcomed the differences—the greater the better—as exciting opportunities for extending my personal life and interests and as a challenge to my adaptability. I also recognized, of course, that, just as a Kurosawa movie is likely to be "influential," so also Japan's role in the world has now become important to the point that anthropologists, sociologists, and economists are probing the reasons for the recent innovativeness (by which

I mean the production of improved or new articles of cultural or economic value) so evident lately (but of long historical precedent).

The three-year-old Rockwell-Hitachi exchange program was set up as a joint, open-ended, precommercial research trial with one person from each organization exchanging each year. Hitachi Corporation has other programs for visiting scientists and engineers (HIVIPS) and welcomes foreign workers, both at its Central Research Laboratory in Tokyo (Chuken, for Chuo Kenkyusho) and at the original Hitachi Research Laboratory (Nikken, for Hitachi Kenkyusho; *Nik-* is the name of the first Chinese character of *Hitachi*) in Hitachi City. For unaccompanied visitors to Nikken, such as myself, Hitachi has opened an "international guest house" in recognition of the well-known difficulties of housing in Japan. For families, company apartments are available. This is a big advantage in the small-town environment, and it's more like the "real" Japan.

At the international guest house, Izumigamori Kokusai Ryo, where Hitachi engineers also reside, male foreigners live in very acceptable single hotel-like rooms. Thus, at buffet-style breakfast and dinner, there is much chance for communication. Amenities include tennis courts, international television, and English-language newspapers, as well as company-sponsored mixers. Foreign families tended to drop by this center for tennis, meals, and other social activities. Foreigners were encouraged to partake in company-related swimming, tennis, and other sporting and cultural events, and most did so. From this perspective, then, the Hitachi Corporation has done much to alleviate the sometimes, by western standards, poor living conditions that may be encountered in Japan. Perhaps other corporations will follow the lead (or are already doing so). Hitachi encourages learning the Japanese language through videotapes, classes, and other methods.

At Hitachi there have been more visitors from Europe thus far than from the United States; I suspect that Europeans' desire to spend time in Japan is greater than Americans', the latter, unfortunately, being perhaps less well prepared by their relatively parochial education. Educated Europeans expect to have to speak at least one language other than their own and thereby become aware of cultural diversity. Americans must expect that cooperation between Japan and Europe (next-door neighbor Russia, too, is a fascinating prospect) will go on whether or not the United States itself is also committed.

By 1989, the new field of high-temperature superconductors was just two years old; since Hitachi and Rockwell (and indeed, it seemed at the time, almost all corporations) were highly interested in this very precommercial development, it was ideal for an exchange subject, with surprisingly minimal legalistic constraints; at the time of this writing the fourth exchange of one person from each company has been initiated for research in this field.

The *Chodendo* (superconductivity) Center of Nikken was founded in the spirit of a longer-term viewpoint, as opposed to the principle of near-term profit. Not presupposing to know the exact scale of, or uses of, a product in the marketplace is a more common phenomenon in the history of technology than the advocates of presumptuous, so-called market analysis may anticipate. The attendant risk is moderated by starting small, "learning by doing," and gaining the know-how for later surprising developments that become inevitable if enough seeds have been planted. (In the United States, to some extent this role is played by the small, entrepreneurial, venture-capital companies.)

Although I had already—perhaps remarkably—met the director of the Hitachi Superconductivity Center, Shin-Pei Matsuda, a few months before at a delightful shipboard dinner party in San Diego Bay as part of a Materials Research Society (MRS) meeting, we had not predetermined the specifics of the research. I presumed only that Hitachi personnel had looked at some of my publications, and I likewise had reviewed the Superconductivity Center literature. In spite of my earlier comments on the need for cultural and, at least, some language interests, I recognized that the technical program must take priority; if that went well, then other matters would, with a little effort, take care of themselves. With poor research accomplishment, the rest would be for nought.

Upon arrival (which was accompanied by unexcelled Japanese hospitality) I gave a seminar of my recent work and was asked what I wanted to do. I suggested that the Hitachi engineers, too, present some of their ideas, so that I could see if we could forge a synthesis of our strengths. At this point I should say that the manner of interactions of this kind will greatly vary with age and experience; Japanese can be expected to react appropriately to seniority (as described repeatedly in this book). Whereas I was expected to lead a joint endeavor (naturally with due regard to Japanese "consensus-seeking" methods), a younger person might expect to be assigned a more narrowly defined role. I was a little surprised that it seemed that we would embark on a project very quickly—I had expected the consensus process and diffusion of ideas to take at least some weeks. In this regard, perhaps the Hitachi Superconductivity Center is unusual; it was newly created, containing more aggressive, fast-moving, and young individuals, many with experience in the United States and Europe, and I felt that they knew what to expect from a westerner. And so, while they were apparently, I thought, adopting a more western mode for me, I was trying to fit a perceived Japanese mode (disintering my long-lost British upbringing?). In any event, after only two days, we arrived at an agreed plan.

At the Superconductivity Center the open-office system, discussed elsewhere in this book, was paralleled by an open-laboratory approach (refreshing, compared with the locked-door fiefdoms of laboratories in the west). Each piece of equipment has a "keeper"; the younger personnel "learn by doing" and "start small," training on the job with the senior personnel, in the well-known Japanese way. Few people enter the system as Ph.D.s; some that do have foreign Ph.D.s. Generally the young people enter straight from the bachelor's level expecting to learn on the job and can be molded in the company "family image" as they anticipate, of course, lifetime tenured employment. This does not mean that a lifetime of research is necessarily anticipated; different roles may be "agreed upon" and expected as an individual moves through the corporation. Doctorates may be acquired later, as D.Sc.s are in Europe. One now understands how the "club" interests at universities predispose and aid students for this later group behavior (see Chapter 7).

This sociotechnical difference between cultures leads naturally to a different outlook on research. In the west, the Ph.D., who controls most exploratory research, has already been groomed for seeking the big, individual, original, conceptually unique, historic "leap forward" as a goal. Not so in Japan, where group harmony will be best preserved by many agreed-upon, even allocated, small steps of achievement, integrating sometimes to a greater end, albeit not with the fireworks, excitement, and Nobel Prizes of the western model. Japanese often comment that they do not distinguish between science and technology as they hear westerners discuss them. This could be the subject of another article, but the process of science seems to be viewed as essentially the same as that of technology. For better or worse, I have yet to meet a Japanese person who has read Karl Popper's work, that of his acolytes, or the later scientists such as Kuhn,

Polanyi, Medawar, and Hallam, and others who have discussed "philosophy" of science from a social or a working scientist's viewpoint (and whose books I recommend). Few working scientists or engineers in the United States have done this either, let alone indulged in Buddhist readings, but it is the sort of interest that the western scientist can bring to Japan, since creativity in general is a fascination of the Japanese. Since it is an active interest of mine, it was a credential that I could bring into play, and therefore, in an English conversation group I was leading at Hitachi, we discussed scientific creativity (indeed, any creativity) as well as many other topics. (Occasionally I was guilty of "lecturing"—not a good idea.) This is just an example of how a foreigner's special strenghs may be appreciated by the group in a Japanese corporation.

My own personal background is in the phenomenology of complex ceramics, including synthesis, sintering, grain growth, and related matters. Ceramic engineering has been a particular focus of interest in Japan within the broader materials realm. The noted Japanese success in this field is attributable to an emphasis on processing, with more heuristic, chemical approaches rather than the more physical, analytical American stance. Hitachi has a strength in the technology of silver-encapsulated tapes, and a peculiarity had been noted in that the silver was surprisingly affecting the behavior of the high-temperature superconductivity (HTSC) in a generally beneficial way. How could we understand the reasons for this? Bearing in mind that I had only six months of initial interaction with the group (although I hoped our association would continue later, and it did), I tried to think of an efficient method of studying this without the hundreds of experiments normally required in a multiparameter ceramic system with up to seven elements and extreme sensitivity to temperature and time, for instance. The particular outcome was that we realized that by placing the unfired 14-cm tape in a constant thermal gradient and afterward examining the lengthwise-polished section in various ways, we could perform essentially 100 temperature runs with each experiment. The details of the work, presented in Japan and in the United States, have been published;[2]

[2]P. E. D. Morgan, M. Okada, T. Matsumoto, and A. Soeta, "A Thermal Gradient Technique for Accelerated Testing of Tl-HTSC (or, for That Matter, Any Ceramic)," in T. Ishiguro and K. Kajimura (eds.), *Advances in Superconductivity II: Proc. 2nd Int. Symp. on Superconductivity, ISS '89, Tsukuba, Japan, November 1989*, Springer-Verlag, Tokyo, pp. 435–38).

the importance is that we well realized the synergy that could be gained by using a Japanese strength, in this case the considerable technology of making the tapes, with the scientific insight of the thermal gradient method, now very practical with the long tapes that were readily available.

"Basic" science is well known to be inspired by technology (for example, thermodynamics by steam engines), and the linkage is something that the Japanese can now expect increasingly to be able to make. The Japanese are looking into the western ideal of leapfrog invention, while the west increasingly scrutinizes, with some trepidation, apparently, the successful *dan-dan* (step-by-step) or *kaizen* (improvement, implying many small, never-ending, steps) approach, since in the west consumers vote with their dollars. Much will be achieved by an eventual synthesis of these creative attitudes.

In my personal interactions I did run into a few situations which were not unexpected, after the reading I had done, but which are worth passing on here. The Japanese don't like to say no; "it is difficult" or "maybe" usually means no. It is much better to ask for suggestions than make requests; demands are out of the question. If a question or request is not answered, don't push, but wait and rephrase the inquiry, give reasons for the request, even wait for a social situation (over *sake*, for instance) to raise the issue. Humor, if you can bring it off, is acceptable. Sometimes, in trying to get an answer to a "difficult" question, you may feel that you have been given the answer to a completely different or unasked question. Let it pass. Assume that language difficulties may have caused this; even if they haven't, this allows the other side a way out. Learn to think of this from the Japanese point of view and always allow time.

Another issue is the *uchi-soto* viewpoint—roughly, but more strongly, "us and them," in English. If, according to plan, you integrate into your group, becoming uchi, you can become soto to another, competing group. Unofficial attempts to build bridges, without the aid of higher authority, say, from the company above, and especially without proper introduction or advance warning, can be viewed with extreme suspicion. A person may claim to be too busy to see you, for example. This is understood to be a put-off, and you would have to devise a plan to ameliorate the situation for the future. This may also occur between you and people in other companies. It may be that you know a person in one group very well but another person you want to meet is in a competing group and, although this second person may share your larger professional interests very closely, nevertheless uchi-soto may be an overriding factor; treading on toes is to be avoided at all costs. I would not overworry about these situations to the point of paralysis, however. Individual temperaments vary much in Japan, as elsewhere.

Once again, this can only be a personal recapitulation of my own experiences, but I hope I may emphasize that while you read (as in other chapters of this book) about Japanese ways, corporate culture, and so forth, you should not be overly inhibited. In the end you must be yourself cast into this role, use your own inner resources in the part so that the production is improved, remember that sincerity, giving, understanding, concern, patience, sensitivity, and warmth are truly international attributes that will be appreciated and reciprocated. Speak with actions, for the Japanese are not ruled by economics first but by human relationships. It is occasionally said that the Japanese are ruled by emotions more than westerners (a surprise to those who think that the Japanese are taught to suppress emotion—yes and no!). A gross generalization would be that, whereas many Americans regard their work as a means to buy the pleasures of "real" life when they are not at work, the Japanese see their work as nearer to the core of what they are—akin to the way in which more creative people everywhere generally see themselves. Perhaps westerners would do well to wonder if the sensibilities of music, poetry, literature, and religion, which we hope illuminate the curiosity of our lives, are not nearer to the surface in Japan (as I believe they are in Europe also). Wonder why the Japanese observe *sakura* (cherry blossom time) with a kind of melancholy; for life, like the beauty of cherry blossoms, is ephemeral (perhaps even an insubstantial dream). Know a little of the differences between the Buddhist and Judeo-Christian views of the world and attempt to probe the more than 2000-year-old Japanese soul or spirit (*kokoro*—a word constantly used in literature), and you will be on the cultural track at least. All of this will give you plenty to do, along with learning the language, while you have those slow startup periods that are often commented upon in Japan.

Many small, unanticipated, pleasures occurred during my stay; for instance, when two of my colleagues from Rockwell who had not been to Japan before came visiting, I was able, with the knowledge I had acquired of the local region through the hospitality of my Japanese hosts, to accompany the visitors to the local gardens and temples. This was a welcome respite from business, and an opportunity to show off some places unlikely to be visited normally, say by tourists. In this way I was able to convey, I hope, some of my own interest in and affection for things Japanese.

On returning (unless, like Lafcadio Hearn, you never return) your experience will be of value in ways difficult to anticipate. One of my personal pleasures is to escort Japanese visitors to the United States to places and events that I now know better they will greatly appreciate and to enjoy the Japanese sense of humor that emerges quite quickly if it is cherished and reciprocated.

In the first year after I returned to the United States, I revisited Japan three times for business, now with added enjoyment resulting from the earlier experience: from the humorous and anecdotal stories I had heard, and from the knowledge of the language and customs I had gained during my sojourn.

A CIVIL ENGINEER'S EXPERIENCE WITH THE JAPANESE CONSTRUCTION INDUSTRY
Charles R. Heidengren

In the present economic environment, Japanese firms, as well as most international organizations, have realized the importance of working with foreign-born engineers. Not only does this policy enhance their ability to work more efficiently in an environment outside their native land, but it also promotes meaningful exchange of technological ideas and information.

My comments are based on more than 10 years of living and working in the Japanese construction industry. Five years were spent with a large private civil engineering consulting firm, and the $5\frac{1}{2}$ years with the engineering and construction division of a large steel manufacturer, where activities included turnkey design-construction projects and general contracting. My activities included business development, technical proposal preparation, prequalification, competitive bidding, and contract administration. I was hired as a specialist.

The construction industry in Japan, like that in the United States, plays a crucial role by providing the structures that house and facilitate virtually all other economic and social activity. In addition, this industry has historically played an important role outside Japan, not only through direct export of goods and services, but also through exercising leadership in opening opportunities for other Japanese businesses and for intellectual exchange that improves international understanding.

The importance of technological leadership is widely recognized in the industry as a key component for enhancement of the competitive position in an increasingly global marketplace. In order to remain competitive in international markets, traditional technological advantages must be maintained and the skills needed for competition must also be developed. It is in this latter area where the foreign-born engineer can play a key role in Japanese industry.

Civil engineering technology in Japan is excellent. Management-related technology is, however, somewhat lacking. In my experience the organization of project team activities, scheduling, and cost control in Japan were not always as vigorous as "western" practices. Computer technology, too, was at one time limited by a lack of suitable software in many disciplines, including construction engineering; this situation is now improving. Japanese civil engineers practicing in private industry tend to rely on their own personal experience or on government manuals.

Specifically, as a foreign-born engineer, one can expect to observe the following deficiencies in engineering practice:

▶ Lack of use of innovative technology by private consulting engineers. More creativity is shown, however, by the large contractors, suppliers, and manufacturers.

▶ Japanese industrial and other standards are used for design criteria, materials specification, and the like even when not the most suitable for overseas locations, because of climatic and topographic conditions, such as in very hot or wet tropical areas.

▶ Management policies are not always very efficient, thus leading to lack of coordination and poor organization of work methods and procedures. Inadequate staff scheduling often results in cost overruns and low profit margin.

▶ Design or contract drawings are sometimes poorly laid out and inadequate in detail.

▶ Most engineers in Japan are "generalists." State-of-the-art technolog-

ical advances may be limited because few specialists are available to perform geotechnical, hydrologic, environmental, or structural work.
▶ Higher costs to the owner may results from a conservative approach to design without innovation or creativity. An ad hoc philosophy leads to setting (and trying to meet) impossible deadlines. There is no time for innovation or creativity; instead, designers rely on design manuals from the government or manufacturers.

What Is Positive

At the top of the list of strengths are the excellent Japanese work ethic and sophisticated designs based on advanced design criteria. Japanese technology in tunneling and earthquake engineering is on a level with the most sophisticated and best in the world.

Contractors and designers in Japan maintain realistic factors of safety in their work. Corners are not cut to save money. Basic research is very often implemented for complex engineering problems, particularly those involving new technology.

Simulated testing, model testing, and computer analysis have accomplished a great deal in applied research leading to development. This work is most often done by manufacturers, contractors, and suppliers.

Japanese company employee training programs are generally good. A foreign engineer can generally expect to be involved in such programs.

Diversity

In Japan, construction investment equals about 16 percent of the gross national product. This compares with only about 10 percent in the United States and other industrialized countries. The previously underdeveloped state of infrastructure in Japan, including parks, highways, bridges, airports, sewer systems, and housing, has accounted for most of this difference. In most western industrialized countries including the United States, construction companies usually function only as builders, whereas in Japan construction firms are involved in architectural design, urban planning, civil engineering, and land development. They are also now at a turning point in their attitudes toward marketing of services.

Foreign-born engineers will observe another important difference between practice in Japan and that in the west. There is a high level of research and development by Japanese general contractors. Large numbers of top researchers working in laboratories furnished with the latest scientific equipment comparable to that found in American universities are working on problems related to new markets. Some of the topics being investigated with keen interest as we approach the twenty-first century include advanced technologies in biochemistry, space development, air pollution, and hazardous and toxic waste disposal.

Foreign-born engineers will find the attitudes, motivation, and capabilities of young civil engineers in Japan to be different from their own, generally speaking. Few seem interested in overseas assignments, because of a preference for the Japanese lifestyle and family goals, and because of problems they would have in overseas countries with two-career households, children's education, and communication. Communication in a second language, both written and oral, is very difficult for many Japanese, not only engineers. Efforts are being made in both the public and private sectors to solve this problem with intensive foreign-language studies.

Most contractors and consulting engineers in the United States encourage young engineers to participate in the activities of professional societies such as ASCE (American Society of Civil Engineers), ASME, and AIA (American Institute of Architects). Professional society meetings, conferences, and continuing education programs are normally considered an important part of career development in western countries. The average Japanese engineer does *not* participate to any great extent in professional activities. Instead, most efforts are devoted to the company's goals.

Future Expectations

In the future, I expect that Japanese practicing engineers will participate more actively in international conferences by presenting technical papers and working on committees. They will be more willing to adopt ideas and developments from the United States, the United Kingdom, France, Germany, and other countries.

Professional attitudes and understanding will also be very important to carrying out the "megaprojects" to be initiated in Japan. An estimated $6.6 trillion (U.S.) will be spent during the 1990s to improve, maintain, and repair the Japanese infrastructure. Professionalism will play a significant role in satisfying the growing need for cooperation between Japanese and foreign engineers.

In the future, there will be more investigation and awareness of differences between Japan and the developing world in such matters as labor, materials, equipment capabilities, environmental conditions, and sociological pressures. There will also be more willingness to share Japanese technological innovations and developments with engineers everywhere.

A EUROPEAN ENGINEER'S VIEW OF INDUSTRY
Dimitrios C. Xyloyiannis

This case study stems from my intermittent involvement with the Japanese people since 1963.

My first encounter with the Japanese was in 1963–64, during repair of a tanker ship at the Hitachi Shipbuilding Company. Later, in 1974–77, I supervised (as a surveyor, just after getting my M.S. in mechanical engineering) the building of 12 bulk-carrier vessels with a capacity of 33,000 deadweight tons at the Kanasashi Shipbuilding Company. During that time, I had to travel extensively to many places in Japan for the shop trials of the ships' main and auxiliary machines.

For the next 12 years I worked as chief engineer for a multinational pharmaceutical company. In 1989 I was transferred to my company's Japanese affiliate as project engineer for the construction of a factory in Saitama and a research institute in Tsukuba.

My experience has taught me that when dealing with the Japanese one has a high probability of working out a good relationship if one considers a Japanese partner to be trustworthy and of at least the same capabilities as oneself. In most cases, I have found that when there is a conflict between a foreigner and a Japanese, the Japanese is right most of the time. Of course, it takes two people to make a conflict, and I do not claim that the Japanese are always right, but at least they are often responsible to a lesser degree. Therefore, we foreign engineers have to look into the cause of any conflict that arises, see where we are wrong, and admit it immediately.

Foreign engineers who plan to work with the Japanese, especially in Japan, must do their homework before they get involved in any Japanese business. I would suggest they they learn as much as possible about Japan and its people: its culture, customs, methods of decision making and follow-through, business values, methodology, and working hours. Socializing with the Japanese and studying their behavior can also enchance business relationships.

I personally found the study of the Japanese language very beneficial. I did not have to master the language, but by trying to learn it I could understand much more easily why my Japanese partners have so much difficulty understanding foreigners.

Making trips to the countryside (small cities such as Fukui and Fukushima) helped me to observe the real Japanese way of life and gave me some insight into the people's behavior. Big-city society is somehow "contaminated," as a Japanese colleague once expressed it to me.

To paraphrase a familiar saying, when in Japan, do as the Japanese do. My experience is that by applying constructive behavior, one has a much greater chance of being successful. Constructive behavior includes everything that will contribute to gains for all parties involved. In Japan the sense of balance is very intense. Therefore, deals and bargains struck on a win-win basis (not I win, you lose) offer the highest possibility for success. In a nutshell, it is wise to do the following:

▶ Encourage your Japanese colleagues to do most of the talking.
▶ Listen carefully and verify understanding.
▶ Avoid comparing cultures (especially criticizing) unless you see some positive similarities. Japanese culture has survived and been successful for centuries, just as it is. By criticizing, one loses ground and to a high degree creates, if not a negative, at least an uncomfortable environment. I would like to emphasize this point, because I have seen many foreign engineers fall into this trap.
▶ Avoid showing off. The Japanese people are very eager to learn and humbly accept that they have learned and can learn from foreigners, but they dislike being lectured.
▶ Be consistent and be a doer. In other words, be sure that what you say, you can do.
▶ Keep a written record of cases in which you have failed or succeeded, and make guidelines that fit the working environment you are in. What works in my case does not necessarily work in your case.

In conclusion, one should consider it dangerous to take one case or a few cases as rules, no matter how authoritative they seem. This is particularly true in an area as complex and controversial as the relatonship between people, from different cultures and sometimes different educational disciplines, engaged in reaching a common goal, especially if they have dissimilar interests.

References

I have read many books on the subject of Japanese business relationships, but here is a list of books I would thoroughly recommend:

Mark Zimmerman, *How to Do Business with the Japanese: A Strategy for Success,* Random House, New York, 1985. Zimmerman presents excellent suggestions for working sucessfully in a Japanese company (Chap. 17, p. 254).

H. Jung, *How to Do Business with the Japanese,* Japan Times, Tokyo, 1986.

Takeo Doi, *The Anatomy of Dependence,* translated by John Bester, Kodansha International, New York, 1982.

A. Young, *The Sogo Shosha,* Charles Tuttle, Tokyo, 1989.

16

Observations on Working in a Japanese University

Michael W. Barnett

INTRODUCTION: THE SETTING AT KYOTO UNIVERSITY

Kyoto Imperial University was founded in 1897, the second university to be established in Japan. The College of Science and Engineering was opened in September of the same year and was separated into two colleges in July 1914. After the Second World War the university was renamed Kyoto University. In 1949 it was reorganized into a four-year university, and in 1953 a graduate school was founded. At present there are 158 chairs (about 550 faculty members) in the faculty of engineering, which is organized into 23 undergraduate and 25 graduate departments, and 4 research laboratories. There are 5930 undergraduate and graduate students enrolled in engineering, of whom fewer than 3 percent are women (this percentage is typical of Japanese engineering departments). The total student population averages around 17,000.

At its founding, Kyoto University was a state-oriented school, its main purpose being to meet the specific needs foreseen by the nation's leaders. Over time, the focus of the university has changed to address more completely the needs of society as a whole, though it has maintained a slight antistate bias in contrast to the University of Tokyo, which, since its founding, has always produced many government bureaucrats. Kyoto University has had a colorful past, having been a center for the Marxist movement before World War II and the Red Army after the war and home to more Nobel Prize winners (four) than any other university in Japan. Kyoto

University is regarded in Japan as an excellent school for engineering and the sciences. In order of prestige among national government schools, Kyoto University is, arguably, number two behind the University of Tokyo.

This chapter summarizes observations I made while working at Kyoto University as both postdoctoral researcher and lecturer in the Department of Environmental and Sanitary Engineering. Perhaps it is obvious that a good knowledge of Japanese society goes a long way toward the understanding of Japanese institutions such as a university. But, because of the greater degree of homogeneity in Japan, there is more congruency among institutions than in American society. Thus, it is a little easier to generalize. Below the surface some things in Japan don't seem to change much over time. Still, when making judgments it is important to be cautious, since the situation is dynamic, especially now that Japan is finding it necessary to take a leading role in international affairs. Economically and otherwise, the international community has been rather accommodating, but now Japan needs to reciprocate. Ensuring a successful transition to a more open society is one of the challenges facing Japan today.

FORM AND SUBSTANCE

On the surface, Japanese universities can appear to be much like western universities. The university's mission in society is clearly defined, the administrative organization chart shows several faculties and institutes overseeing various university activities, classes appear to be western-style, and research laboratories are much like those in the United States. Yet, close inspection reveals many differences. The university structure is rigid, there is little movement of persons or ideas horizontally, the promotion system is inflexible, the quality of lectures varies quite a bit, and the behavior of both teachers and students in and out of the classroom is completely different. In fact, the contrast is so great that it is tempting to conclude that many activities are rituals performed for no other reason than that it is written somewhere that it must be done this way. Outward appearance is not a good indicator of what goes on inside, and there are significant problems; however, in many ways the system works. The education system, of which the university is only a part, does develop competent engineers and scientists, but the manner in which this is achieved differs considerably from the way it is done in the west.

It is well known that since the war the interests of the economic producers have been highest on this nation's priority list. It has also been noted that Japanese industries in their recruiting practices place a lower premium on specialized skills than do western industries, and these practices have in turn affected the nature of university education in Japan.[1] Emphasis on the economy, combined with the need to supply industry with persons having "latent ability," has cast universities, especially engineering departments, in a specific industry support role and left less room for purely academic pursuits. American universities stress specialized, individual professional development and creative, scholarly research, with companies largely being forced to accept what they get. In contrast, Japanese universities emphasize cooperation, coordinated development of needed skills and technology and are sensitive to the demands of business and industry. Industry, for its part, has shown a willingness to shoulder much of the responsibility for training and thus is an important part of the education system in Japan. This appears to be one of the strengths of the system.

University education stands in sharp contrast to primary and secondary schooling in Japan. The characteristics of basic education in Japan have been documented, as have those of Japanese universities.[2] One problem in the university concerns the students. Some, even at the master's level, devote more time to extracurricular activities than would normally be permitted at an American university. Young Japanese know that acceptance to the right university, by passing a series of rigorous examinations, will both ensure them a good job and allow them four or more years of free time to devote to club activities as well as, for the first time since early childhood, enjoying life. This often becomes a consideration when deciding upon continuing education to the graduate level. A large fraction of students take the entrance exams three or four times (exams are given once a year by each university), since many students apply for admission to more than one university. There is little or no movement laterally between

[1]Robert C. Christopher, *The Japanese Mind: The Goliath Explained,* Fawcett Columbine, New York, 1984 (also Ballantine, New York, 1984), p. 93.
[2]Edwin O. Reischauer, *Japan: The Story of a Nation,* 4th ed., McGraw-Hill, New York, 1989, pp. 186–202.

departments, and students who pass the examinations must decide early which career to pursue. At Kyoto University, freshmen are accepted into a specific faculty and, in the case of engineering, a specific department. The system is rigid, but on the other hand, it can be argued that the average Japanese student has a solid foundation in basic learning skills, having struggled for many years in elementary and secondary school. Skills are uniform among all groups of students majoring in engineering and the educational level of entering freshmen is probably higher than in the United States. There is also uniformity in the sense that everyone seems to know the same things. Rote learning in primary school is heavily reinforced, for example, as needed to master the complex writing system. Individualistic tendencies, including certain creative activities, are generally frowned upon in the university as in Japanese society as a whole. Conformity to the group is considered best.

The manner in which lectures are conducted in a Japanese university often shocks outside observers. In the social hierarchy a professor is held in high regard and it is impolite to question his or her authority. Students are nonaggressive and reserved in their behavior. Few, if any, questions are asked in the classroom. The teacher prepares notes for distribution and delivers lectures uninterrupted. It is difficult to get Japanese professors to lecture to groups of foreign students, since Japanese professors are uncomfortable with the confrontational character of western-style teaching, where student and teacher may challenge one another. There is much talk about "challenging the student to learn," but here little can be done without some cooperation from the students. There is a kind of nongrading grading system in which only the worst students are given failing grades and essentially all undergraduate students finish in four years; master course students, in two years.

Administratively, the department is a blend of western-style rules and procedures and Japanese-brand bureaucracy and decision making. Weekly meetings are held to discuss department business, but decision making calls for many behind-the-scenes discussions with affected groups. Important decisions are effectively made before being presented to others at the faculty meeting. Junior members defer judgment to full professors, and a professor's judgment is never questioned. In a very real sense the department is like a family. As is typical in Japan, 80 to 90 or more percent of the faculty members are previous graduates of the department in which they currently work. A hierarchy of prestige is maintained thus; for example, much attention is given to the ordering of names on official documents. Many activities are tightly controlled by the Ministry of Education (*Monbusho*). Changes in personnel, such as promotions to professor, or curricula, such as addition of a new class, become political issues and may take months or years. The degree of control exerted 'is not always in proportion to the amount of funding provided by Monbusho. A professor (the person responsible for obtaining and distributing funds through his or her chair) who chooses to obtain financial support from industry or business may, if successful, fund most of the chair's research activities from these sources. Funds for overhead, such as building upkeep or remodeling, and other indirect costs are insufficient, and university salaries as a whole are low compared with those in American universities. Outside consulting (by employees of government institutions) is not permitted, but with the blessing of one's professor, honoraria can be collected from companies for giving lectures or technical advice.

The quality of a given chair is highly dependent on the quality of the professor in the chair, since this person, like the head of a family, is the primary funding source and decision maker (sometimes including decisions regarding employment of graduating students through the chair). The system permits qualified individuals to move to the top but also drags with it some poorly motivated others. The essential policies of those wanting to advance in the system are to do what you are told, never question your superiors, and persevere. Eventually, the system will reward those who are patient. Good professors know this and grant promising individuals the funds and freedom to do their work. Bad professors take advantage of the situation by dictating to their staff and students, stifling creativity, or just doing nothing. The system permits all these behaviors. It is possible for the complexion of a department to change considerably with personnel changes or the mandatory retirement of a professor (at age 63 for Kyoto University). Power can become centralized, and because professors are held in such high regard, one professor can have a great influence on work in an entire discipline. The professor is always listed as coauthor on publications, regardless of actual contribution.

Too often appearances are weighted more heavily than substance, but the overall goal of excellence in education and research is not forgotten, though it is sometimes set aside. The working environment is generally good, though the facilities are somewhat old and cramped, and as in American universities there is freedom in setting one's schedule. If a niche can be found in the group, effective teaching and research can be accomplished. It is wise to invest some time finding this niche before deciding to jump into the system.

AT THE ALTAR OF INTERNATIONALIZATION

The president's foreword in the 1988–89 Kyoto University Bulletin concerns international academic relations. The president is well known for his activities in this area and is a great supporter of exchange with foreign countries. His policies have been successful in creating a good environment for international cooperative exchange at Kyoto University.

It is possible to hear something about "internationalization" several times during a normal day in Japan. It is a word that has been on the lips of every Japanese for several years, having been placed foremost in the public consciousness by government leaders who saw the need for greater understanding of other countries and cultures. It can be witnessed in many forms, such as the increased availability and diversity of imported goods (at a price) and the adoption of foreign ideas. Kyoto and other large cities in Japan are very international places—more than many American cities, for example, those in the Midwest.

There are two faces to internationalization, one good, one bad. Internationalization has been successful in Japan, though there is still resistance to change. As mentioned previously, the university is quite internationalized and many foreign students, researchers, and visiting scholars can be found on campus. Prominent scholars often visit the university to make presentations. Though the university environment may not be typical of Japan as a whole, there are genuine feelings toward and desire to learn from foreign visitors. It is possible to have candid discussions with colleagues on most topics, and the majority understand the difficulties of adapting to a new culture. Japanese are good hosts and will typically go to great lengths to help you if you have problems.

The dark side of internationalization hides a significant problem in Japan, namely, strong feelings of separateness and uniqueness. Of course, it is good to feel that one's culture is unique; however, in Japan these feelings are intense, as though the uniqueness itself were unique. Japan, willingly, feels separate from the rest of the world; thus, the world feels separate from Japan and reacts with distrust.[3] Japanese are being pulled in different directions, finding it necessary to be both open, that is, to be internationalized, and closed, that is, to maintain their Japanese-ness. The tension has resulted in a degree of neurosis in Japan. At the university, this is felt in several ways. Students, researchers, visiting professors, and others are invited without adequate consideration for work they might conduct or how they might fit into the group, papers for many "international" symposia in Japan are solicited but never fully appreciated or even read, and qualified technical professionals find themselves occupying slots as token foreigners and English teachers. One sometimes feels as though one were being sacrificed in the name of internationalization. Many problems are related to language and cultural differences, and the foreign visitor can lessen such problems through his or her own study. Still, it is necessary to be wary of those who are not truly being open or sensitive to your professional needs. Fortunately, many in the university are aware of the problem. Since they must work within a system that is not always accommodating, it is important to be patient, but most problems can be worked out (in a Japanese manner) to the satisfaction of both parties.

RESEARCH AND STUDY IN JAPAN: ADVICE TO VISITING PROFESSIONALS, RESEARCHERS, AND STUDENTS _____

The best advice to give to individuals interested in coming to Japan for work or study is to plan ahead. It is difficult to "parachute" into Japan and conduct meaningful work without a certain degree of luck. The preferred route is a preplanned one which is the result of both sides' having been familiarized with each other's work, perhaps as a result of a previous association, and having discussed specific goals and objectives of the work to be done. I list here a few specific points with the hope that the reader may find them useful in planning work or study in Japan:

[3]Edwin O. Reischauer Center for East Asian Studies, Paul H. Nitze School of Advanced International Studies, Johns Hopkins University, *The United States and Japan in 1990: A New World Environment, New Questions*, Japan Times, Tokyo, 1990, p. 89.

▶ Be patient. Things take time to develop. Make an effort to become part of the work group, and have work you can do on your own so that, if you must wait, your time won't be wasted. It may take some time to coordinate your activities with those of your laboratory, and you may begin to feel frustrated. Unfortunately, some professors do not make an effort to assimilate foreign visitors or facilitate their study or research. Avoid these professors. Some visitors have been known to give up entirely and devote their energies to other activities. If you find it impossible to do work in your field, spend your time studying the language and culture.

▶ To the degree possible, try to learn the Japanese language, preferably by completing course work before coming to Japan. Effective language teaching methods are generally not in widespread use in Japan, and language schools are expensive. Surprisingly, some experienced students of Japanese think it is difficult to learn Japanese in Japan. While in Japan, budget some time for language study and remember that a high degree of cultural sensitivity is vital to effective communication. It is essential to develop, in any language, a good rapport with your work group. Go to parties and social functions, resist the temptation to separate from the group, and maintain a positive attitude.

▶ Japan is expensive and pay scales are generally lower. Also, housing is a serious problem. Your standard of living may be lower than in your home country. Be ready to make some sacrifices. If you choose a Japanese style of living, including the food you eat, it will be easier.

▶ It is a practical reality that, in and out of the university, full professors are powerful people; thus, you should make an effort to cooperate fully with them. Try to understand and work within the system. Don't make unreasonable demands.

SUMMARY

The experience of research or study at Japanese universities can be either good or bad depending on the adaptability of the foreign visitor and that of the laboratory in which he or she decides to conduct work. Universities are only a part of the educational system in Japan, which includes primary and secondary schools as well as business and industry. There is opportunity at Japanese universities for effective research and study, but foreign visitors should take care in planning their stay. The Japanese way of doing things may seem strange at first, but it does have many advantages. Japanese are culturally conditioned for hard work, sacrifice, and perseverance and prefer cooperation over conflict. Yet, the system is rigid, creative tendencies are suppressed, and society is still not completely open to outsiders. Internationalization is having a positive effect, and Japan has certainly shown itself to be adaptable. There is much promise in cooperative work with Japan; it is our responsibility as guests to do our best to understand the system in order to facilitate cooperative efforts and strengthen our ties with this important member of the global community.

I would like to thank my friends and colleagues who took the time to comment on this chapter. Dr. Masafumi Goto of Kajima Corporation–Marine Biotechnology Institute Co., Ltd., Dr. Bernard B. Siman of Jardine Fleming Securities, Ltd., and Mr. Paul Driscoll of the Kyoto University Research Center for Biomedical Engineering were especially helpful. Thanks to Professor Masakatsu Hiraoka and his staff in the Department of Environmental and Sanitary Engineering at Kyoto University. Finally, I would like to thank Dr. Charles W. Knisely for introducing me to Professor Wataru Nakayama of the ASME Japan Chapter, and Dr. Hiroshi Honda for giving me the opportunity to make a contribution to this book.

References

James Fallows, *More Like US*, Houghton-Mifflin, Boston, 1990.

Ivan P. Hall, "Organizational Paralysis: The Case of Todai," in Ezra F. Vogel, (ed.), *Modern Japanese Organization and Decision Making*, University of California Press, Berkeley, 1975.

Karel van Wolferen, *The Enigma of Japanese Power*, Knopf, New York, 1989 (also Random House/Vintage, New York, 1989).

17

Establishing Firms in Japan

INTERCULTURAL PROBLEMS
Ghassem Zarbi

An entrepreneur wanting to enter a sophisticated market such as Japan needs to make a careful study before making any substantial commitments. The most popular form of business organization open to foreign investors in Japan is the *limited stock company,* whose principal characteristics are in many ways similar to those of an American corporation. To establish a limited stock company, a minimum of seven shareholders are required. They need not be Japanese citizens or residents, but all must hold at least one share in the corporation.

I worked for Japanese manufacturers for a few years before I committed myself to establishing my own computer training center. I felt the need to improve my language ability for good communication and to familiarize myself with the ways business is handled in Japan. Since my wife is Japanese, it was much easier for me to start my own office; I received assistance and advice from my wife and her family.

I have closely studied shop-floor skills and industrial relations in Japan for many years. I have also been directly involved with scientific and technological research work conducted by universities, special research institutions, and private enterprises, all of which play important roles in industry.

The foreign investor has no choice but to adopt Japanese practices as they are, albeit in slightly modified form. Because of their unfamiliarity with the Japanese market and language, foreigners have rarely attempted to undertake direct sales operations without the assistance of an independent Japanese representative.

An analysis of the Japanese personality—identifying the ways in which the Japanese really are different from other nations in doing business and carrying out other activities—would ease many matters for foreigners who wish to live and establish a firm in Japan. Perhaps one of the ways of beginning to understand the Japanese, their nation, their customs, and their concepts about life is to imagine going through the same basic learning process about things, country, and people that the Japanese go through in childhood.

There are some characteristics of the labor force unique to Japan that a foreign investor must be aware of. The highly educated Japanese labor force has been put to effective use by industry in a system guaranteeing lifetime employment and wages based on seniority. Also, Japanese attach importance to their group, which in the case of workers is their company; they cooperate closely in group projects with little expectation of immediate reward.

Engineers fresh out of the university are assigned to jobs on the production line. The hands-on knowledge these people acquire of actual production processes and product design and development permits them to make more meaningful contributions in their fields.

The sociological climate in Japan, unlike that in Europe and to a lesser extent unlike that in America, provides an open road to higher rank and many opportunities for individual mobility. This promotion incentive in the business world encourages self-improvement and is the reason for the high morale seen in most Japanese offices and factories.

There are, of course, different ways of conducting business in Japan for the foreign investor. Direct sales operations at both the wholesale and the retail level through subsidiaries or branches, when permitted by the government, may be quite feasible, provided that the foreign seller is able to hire competent personnel to assist in making the sales presentation.

It is becoming imperative, more and more, for the Japanese not only to know more about foreign countries but also to take every possible opportunity to assist people everywhere in obtaining a broader and deeper understanding of Japan. There are quite a few rich sources of information concerning the establishment of a firm by either Japanese or foreign investors. The Japanese government publishes reports on both its short-term and its long-term plans for industrial development, social welfare, labor relations, and related major concerns of business. (These plans do not have much in common with the government planning found in most other nations).

If a foreign establishment has to employ Japanese labor and work with it, it is essential to have some idea of the makeup of the labor unions and how they look after the welfare of the labor force. The Japanese labor system at present is based upon the worker's union, but efforts are now being made out of necessity to adopt some of the beneficial characteristics of the industrial union; these adjustments in time of need are one of the unique qualities of the Japanese labor movement. For example, long-term employment and steady raises over a long period have proved essential elements of the working environment for building an intellectually skilled labor force. With changes in the employment climate, unemployment benefits—one of the features of an industrial union system—are being instituted to ensure the welfare of employees and their families.

There are alternatives to opening a full-blown manufacturing operation in Japan—such as opening a branch office of an overseas company, which for legal purposes will be treated the same as any Japanese company. It should be borne in mind that even establishing a simple sales office brings with it the problems of management, office rent, staffing, housing, expense accounts, and the recruiting and training of a sales force.

One of the keys to success in doing business is advertising. Japan is ideally suited to mass media advertising. Other important channels are visits to prospective client companies to establish direct contacts and to introduce your firm's line of work, and direct mailing. The expense of advertising is often discouraging to a foreign investor. However, it should be understood that in Japan manufacturers are compelled to make constant efforts to develop new products and improve existing ones. Without continuing advertising efforts manufacturers are not able to maintain a fresh image.

The decision to tie in with a general trading company should be based on a careful investigation of the Japanese distribution channel for the product involved. The Japanese government permits the establishment of wholly foreign-owned subsidiaries for foreign investors who wish to establish facilities in Japan to import, warehouse, market, and service their own products, within the limits set by the government for the product type.

I have found while working closely with Japanese manufacturers and importers of foreign products that, rather than use a trading company for distribution, a foreign firm may occasionally appoint a Japanese manufacturer as an agent. It seems to be more beneficial, since the firm gains access to the distribution network and market reputation of an already established company.

There is yet another unusual distribution method used in Japan. In some areas of industry, for example, car industries and steel mills, large Japanese manufacturers deal with a number of small companies that function as their subcontractors. This helps establish the product image more rapidly and reduces the tension on the distribution line. It should be noted that direct contact between Japanese end users and *foreign* producers is not common. Most foreign producers prefer to go through intermediaries.

Foreign investors must investigate their own market and should attempt to find out where their product should be sold, in what quantities, whether there are similar products already on the market, and which lines of distribution, advertising, and approach should be adopted. Sometimes it is necessary to modify the product to suit the tastes and buying habits of the Japanese.

Because of language problems or insufficient financial resources, many average small Japanese manufacturers are not able to establish a trading line with foreign countries. These manufacturers might be approached for the establishment of reciprocal distribution channels if the line of products is appropriate.

There are reports available on the different Japanese industries prepared by the U.S. Foreign Service; intensive information on the Japanese market and trading processes can also be obtained through the Japan External Trade Organization (JETRO). Besides these, there are many market research firms in Japan which can provide you with desired information about marketing, advertising, investing, market demands, legal procedures for export and import, transportation, tax policy, the labor force, ideal locations, and best-suited distribution channels.

AN ENGINEERING CONSULTING FIRM _____
Stephen A. Hann

Most discussions about foreign business opportunities in Japan fall into two categories. Individuals who have started businesses in Japan that have succeeded maintain that Japan is the most open and competitive market in the world. However, those who have started businesses in Japan that have failed complain that Japan is a market closed to all foreigners and monopolized by a handful of insiders. Since my consulting practice in Japan is only $1\frac{1}{2}$ years old, I belong to neither of these groups. This section is my attempt to describe my observations in Japan as objectively as possible.

My discussion is restricted to what I have actually seen during my efforts in building a mechanical engineering consulting service in Japan. This section has three parts: my impression of the business climate in Japan for engineering consulting; a report on how and why my consulting practice was started in Japan; and a list of 10 of the problems that a business will face in setting up in Japan. It is my intention that the third section be the most useful and interesting. This is information that I did not have when I started business in Japan. Had I known these things, I would have structured my business differently and planned it a lot more carefully. Furthermore, I do not discuss the immense rewards for building a successful business in Japan. This is because the required commitment of time, money, and perseverance to achieve anything there is a subject that must be thoroughly understood by anyone with any business ambitions in Japan.

It is my opinion that simply earning a lot of money in a short period of time is not a good enough reason for moving to Japan. For one thing, I haven't seen or heard of anyone accomplishing this. What I have seen is people grinding out a living through what, over a number of years, sometimes develops into a successful business. I want everyone reading this section to understand and concentrate on the challenge of breaking into the Japanese market. The business practices that impede a new venture that is getting started also work to ensure success for those companies that do establish themselves. It should also be kept in mind that only a small percentage of businesses, even larger and well-financed ones, survive the first few years.

I highly recommend two books that will be useful to anyone who is interested in starting any kind of business in Japan. A book that directly addresses the concerns of small businesses is *Setting Up and Operating a Business in Japan*, by Helen Thian (second edition, Tuttle, Tokyo, 1989, ISBN 0-8048-1544-5). While it discusses many items more of interest to importers of consumer goods than to engineers, it is a good overview of the potential rewards and of certain problems that are involved in setting up a business in Japan. Another view can be found in *Competing in Japan: Make It Here You Can Make It Anywhere!* by P. Reed Maurer (Japan Times, Tokyo, 1989, ISBN 4-7890-0486-4). Maurer's contention is that all of the difficulties of doing business in Japan are really opportunities and that Japan is the ideal marketplace.

Opportunities for Engineering Consulting in Japan

While it is true that the Japanese distribution system is resistant to out-siders, I have seen absolutely no discrimination as a foreigner trying to sell services to Japanese firms. This is because the resistance extends to all outsiders, including the Japanese themselves. A Japanese engineer trying to start a business in Japan would face every obstacle (except visa prob-lems) that a foreigner does. In fact, foreign engineers are probably treated a little better and have somewhat fewer cultural problems than native en-gineers. I say this on the basis of what I have seen while trying to build a business that does not directly compete with established companies in Ja-pan; if I had been trying to directly import and distribute foreign cars or computers, my observations might not be the same!

There are relatively few small engineering consulting firms in Japan. The reasons for this are unrelated to the amount of work available and the need for their services. In fact, I am not sure why such a large market has not attracted many more native engineers and firms. For whatever reasons, there is a large, virtually untapped market in mechanical engineering ser-vices. There is a real shortage of engineers who are highly skilled in me-chanical computer-aided engineering (MCAE) software—software using finite elements, solids modeling, and nonlinear rigid-body dynamics—and who can consult on site. Currently, American and European consultants are filling this void, but they are good only for full-time on-site contracts. They come to Japan for the length of the contract—typically a week to six months—work full-time on site, and then leave. If a company wants an on-site consultant only part-time, these foreigners can do nothing for it. A Japanese MCAE distributor may want a consultant available for infrequent but important sales calls and benchmarks, but contractors cannot travel to and from Japan in such a chaotic fashion.

There is an opportunity for foreign engineers who are willing to move to and live in Japan. They must be willing to travel within Japan and work on site. This is a tall order. Most engineering R&D is not centralized in Tokyo but spread out all over the country. On many on-site contracts I spend as much time traveling as I do working. This is not a financial problem; the client companies expect their guests to be traveling from Tokyo and are willing to build the time and costs of travel into the consulting contracts. The problem is the wear and tear on the consultant, but traveling is an important part of providing consulting services and cannot be ignored.

While there are other opportunities for foreign engineers in Japan, on-site consulting and distributor support are the areas of most interest to me. Because of the lack of competition and my ability to meet the needs of the clients, there has been no reason for me to try any other fields. The rest of this discussion will be oriented toward this business.

Establishing a Consulting Business

I first visited Japan in 1987 for a total of three months. I was working in the consulting group of an MCAE software company. Until then most of my projects had been done in the home office and my major customers were local firms. This was my first overseas trip, and it looked like an isolated contract that would lead to no other new business. However, when I arrived in Japan I had a chance to talk to some of our software customers who wanted consulting services but did not want to bring a consultant over from the United States. They wanted to deal with a local firm. The local software distributors had no interest in going into the consulting business. After returning to the United States, I tried to talk my company into sending me over to Japan to start a consulting and distributor services branch office. The company was not interested. The timing was extremely fortunate for me. I had recently decided to quit my job and start my own consulting business and was trying to decide how to find my first contracts. There were ethical considerations involved in contacting former clients, and on the other hand I had no idea of how to sell consulting services to complete strangers. Now, here was a market asking for just the services that I wanted to provide and my company had turned it down. It seemed like the perfect situation for starting a new business. In September 1988 I called my new prospective customers in Japan and tried to get some contracts. The client companies wanted my services but would not commit themselves to contracts until I moved to Japan. The software distributors were a little more cooperative and were willing to give me a retainer to support their software sales and act as distributors for my consulting services. I picked a sole distributor over the Christmas vacation and moved to Japan in March 1989. The whole process took about six months from my first serious inquiry to finally moving to Japan.

It was a real shock to find out how slowly things moved once I arrived. The retainer from the distributor was what initially kept me in business. By the end of 1989 I had billed only 30 days of consulting to clients. The business finally started arriving in January 1990. Since then I have been busy except for three bad months. It looks as if my business is reasonably well established and it is now time to start looking for longer-term and larger contracts. I am also currently looking for my first new employee. To avoid visa problems and to help me with language problem I have decided to hire a native Japanese engineer.

Ten Problems in Setting Up a Small Business in Japan

I have learned a great deal about running a small business in Japan. Here are some items that anyone considering setting up a small business there should look into further:

1. *Setup costs are extraordinarily high in Japan.* While this is the standard advice that foreigners receive when they move to Japan, I was surprised to find out just *how* expensive. A reliable estimate is that everything, except for office space, costs three times what it does in the United States. Office space in Tokyo is available only to well-financed, large companies. Office space within an hour's train ride of Tokyo is still out of reach for small businesses. Forgetting the astronomically high monthly rents, renting office space requires a guarantor (a small company cannot guarantee itself) and on the or-

der of one to two years' rent as deposit (sometimes not refunded). I work on site, at my apartment, and at my distributor's office in order to avoid leasing a separate office. I also live outside of the 23 wards of Tokyo. Anyone who has not budgeted for these costs doesn't stand a realistic chance of succeeding in business in Japan.

2. *Everything takes more time in Japan.* While this is also standard advice for newcomers, it is also accurate. I thought that I had large contracts waiting for me when I arrived in Japan. That was not the case. The work that I actually received from my previous correspondence was about 15 percent of what was promised and it took three months for it to start. Anyone who comes to Japan with less than six months' living expenses (one year would be better) is asking for failure.

3. *American registration is important.* I was surprised how relieved prospective customers were when they found out that my business was incorporated in the United States. Apparently, they are not geared to working with individuals; to do business with the Japanese, it is necessary to be representing a corporation.

4. *There will be visa problems.* The Japanese government is not geared to admitting the employees of foreign small businesses. Since I was not employed by a major multinational corporation, my application for a three-year business visa was rejected without even being considered. I must renew my working visa annually. Furthermore, the only reason I was granted a working visa is that my distributor (a subsidiary of the largest media corporation in Japan) acted as my guarantor. A small foreign business cannot act as a guarantor for its employees. All foreign businesses have some sort of visa problems in Japan.

5. *Discounting does not increase business.* There is no way to get contracts any faster than just waiting. Japanese companies expect to pay full price for services, and offering a discount will not move them to you any faster.

6. *It is impossible to sell goods or services without a distributor.* Major Japanese companies will not deal directly with a small foreign business. They want to deal with a Japanese distributor in whom they have confidence. Trying to start an engineering consulting business in Japan without a local distributor is suicidal.

7. *Choosing a distributor is the most important decision that a small business makes.* I cannot find my own contracts; only my distributor can find them. I am dependent on my distributor for guarantees for my apartment and my working visa. Any increase in my business is more a sign of how actively my distributor is looking for consulting contracts than anything that I have done. A small business that chooses the wrong distributor starts out with two strikes against it.

8. *Customers are extremely demanding with service companies.* It is amazing how many meetings I am asked to attend and reports I am asked to write, sometimes on short notice. In Japan, the MCAE software distributors perform benchmark problems free that an American or European company would treat as consulting and charge for. While we all charge a great deal for our services, we earn it with the many services that we provide.

9. *It is almost impossible to hire Japanese engineers.* Now that my business is growing I am looking for Japanese employees. There is an incredible shortage of engineers in Japan, and the engineers that are there would much rather work for a large and established Japanese firm. There is a chance that I might not be able to find anyone.

10. *Not speaking Japanese is a problem but not an insurmountable barrier.* I am working with English software with associated English manuals. Most of the engineers in Japan read and write fairly well in English and some also speak it rather well. Not learning the language has been my greatest failing in Japan, and I recommend that others not make the same mistake.

In closing, I hope that anyone with an interest in starting an engineering consulting practice in Japan does not think that this section exaggerates the problems in opening a new business there. These problems are some of the reasons so many new businesses, both domestic and foreign, fail in Japan. If anyone feels confident about beating the odds, Japan is a good place to start a mechanical engineering consulting practice.

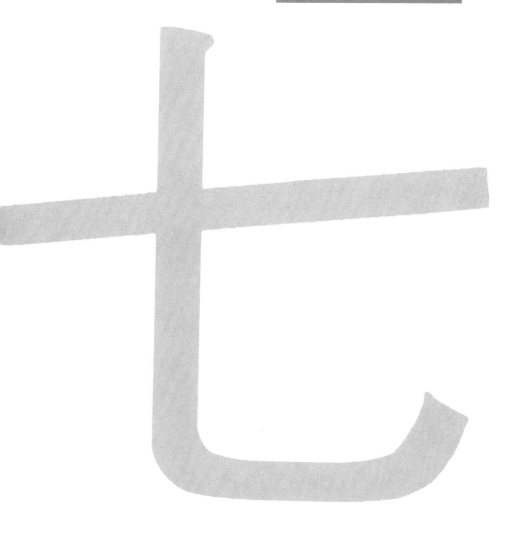

P A R T

7

Conclusions

18

Career Opportunities after Japan

Jon R. Elmendorf

INTRODUCTION

There has been much discussion in the business press and various trade journals over the past few years of the importance of international experience to the growth and competence of American management in today's global marketplace. In some firms an international assignment, if not a prerequisite, is an important factor in becoming identified as a "fast-track" management candidate.

This chapter is presented from the point of view of the engineer as manager and businessperson. While an opportunity to work in Japan is a useful learning experience for those who prefer to focus on pure engineering, my experience and knowledge come from managing an engineering organization. My remarks are also more relevant to the employee of a medium- to large-size American-based company than one interested in working for a Japanese company or establishing a business in Japan.

For the engineer, Japan in particular offers many advantages as an international assignment. Japan is prominent not only because of its recently evident economic power, but more important, because Japanese companies and engineers are very active in the development and introduction of leading-edge technologies in a wide range of industries. Further, living in Japan exposes one to the admirable Japanese disciplines of attention to detail, customer service, diligence, and patience. Such attributes and behaviors will serve the returnee well no matter what his or her chosen career path.

In fairness, there are also some risks associated with a Japanese assignment. The greatest of these is the out of sight, out of mind syndrome. People in foreign assignments can become part of the "lost patrol" when promotional or other career growth opportunities become available in the home office. Also, in spite of all the high-technology activity in Japan, the language and business practices are difficult to learn, and communication difficulties may make benefiting from immersion in this high-tech stream difficult. Finally, depending on one's situation, there is a risk of being identified as "our man in Japan." This appellation can be good or bad depending on one's personal objectives.

PERSONAL EXPERIENCE

I began traveling to Japan on business in 1973. My first stay was for six weeks in a remote area on the Japan Sea side of Honshu to supervise a major inspection activity at a nuclear power plant. Over the next eight years I averaged about four trips per year to Japan for various reasons. As a result of these frequent business trips, I developed many personal relationships among my company's Japanese customers and associates.

Largely as a result of these relationships, I was asked in 1981 to head my company's office in Kobe. The original assignment was two years, but that was quickly stretched to three in Kobe and then one more in our Tokyo office. Thus, in 1985, I found myself in Tokyo, with four years under my belt in Japan and looking for a job back in the United States with the parent company.

At this point it became clear that no one in the parent organization had given any careful thought to my repatriation, or to how my experience in Japan might be used to some advantage. Fortunately, I had a very close and long-standing personal and professional relationship with an executive in another division of my company. I contacted him, explained my situation, and, within a month, was able to choose among four very attractive career opportunities.

The one I chose was as the manager of a newly acquired start-up manufacturing company in southern California. Even though I had no prior hands-on manufacturing experience, my years in Japan had brought me frequent contact with many Japanese manufacturers. Through them I had the opportunity to discuss and study the concepts and application of statistical process quality control in a manufacturing environment. Thus, in the end, my Japan experience proved very valuable to both me and my company.

CONCLUSIONS AND RECOMMENDATIONS

The truth is that very few people are assigned to Japan from companies that have well-thought-out personnel development plans for repatriating their employees. If you are one of those few, you will be very fortunate. In addition, most companies do not have clear policies or plans for taking advantage of the rich experiences and new skills that a tour in Japan provides.

The single most important thing anyone coming to Japan, or to any foreign assignment, can do is to begin planning for repatriation as soon as possible, even before leaving the United States. You should identify one or more mentors in the parent organization, and regularly communicate with them no matter what the reporting relationship might be or become during your time away from home. Keep your mentor apprised of your progress, problems, and career interests, so that when the time comes to return to the fold, you will have someone on your side.

In the end, as is true for many things in life, the benefits from a Japanese assignment are the direct result of one's own efforts. I've known people who have spent a decade in Japan without being touched by it in any way. Conversely, an assignment as brief as my first six week trip in 1973 can be a true growth experience. It's up to you.

19

Summary Remarks

Raymond C. Vonderau

The authors of this book have illustrated among them that professional employment opportunities for foreign-born engineers exist in many, if not most, Japanese companies today. But although this guidebook has been oriented primarily toward graduates in engineering, it is apparent that opportunities for people trained in other professions also exist in Japan.

The remarks and experience of the authors should be helpful for new graduates in planning their career paths to include consideration of employment in Japan. The authors also emphasize the value of the specialized experience of senior engineers in filling the specific needs of Japanese companies.

To provide a representative cross section of the past hiring practices and future hiring plans of Japanese companies, a questionnaire was sent to the head offices of 46 companies in Japan. A response from more than half of these companies reflects a high interest in hiring foreign-born engineers. All of the responding companies indicated that they will hire foreign-born engineers within the next five years. The number of respondents that do not now hire foreign-born engineerings is only 15 percent, which is surprisingly low.

The number of foreign engineers employed by each company varies from a minimum of 1 to a maximum of 40, the average of all companies being 12. The job function of these engineers is predominantly in staff support, research, and product design areas. The employment period of foreign-born engineers in Japan is usually not more than five years, and 80 percent have stayed in Japan for two to five years.

For engineers interested in preparing to work in Japan in the future, the following ranking of engineering disciplines indicates the current hiring plans of Japanese companies:

Rank	Engineering Discipline
1	Mechanical engineering and electronic and computer engineering
2	Electrical engineering and civil engineering
3	Chemical engineering
4	Project engineering
5	Legal services for engineering businesses

Japanese companies prefer to hire prospective professionals on the basis of recommendations of another person, usually from one of their affiliate companies. Slightly over 50 percent of the responding companies hire engineers for assignments in Japan on recommendations from their foreign affiliate companies. The next most preferred means of contacting prospective employees is through university placement offices (25 percent), and the least used means are through advertising in trade journals (13 percent) and through employment search firms (8 percent).

The reason foreign-born engineers are sought after to work in Japan is primarily to obtain their specialized technical knowledge. Slightly fewer than half are currently employed for this reason, which will give encouragement to the more experienced foreign-born engineers desiring to work in Japan. Only 20 percent of the Japanese companies indicated a shortage of native engineers as the reason they hired foreign engineers, and 25 percent indicated that the globalization of their companies required the use of foreign engineers. The possession of product or process knowledge and language skills is an essential asset in making a contribution to these global companies.

The survey showed a trend for new graduates to be hired by a Japanese affiliate company in a foreign country and, after a time, to be sent to the head office in Japan for a training period of two to three years. This appears to be an excellent means of achieving an assignment in Japan for new graduates who have a career interest in the engineering disciplines previously identified.

In conclusion, let me observe that it is the wish of all the authors that this guidebook will provide useful information for those engineers seeking a rewarding and unique opportunity to enhance their careers through employment in a Japanese company. The age of a global economy has arrived, and the future of the "international engineer" has never been more bright.

Note on the Origins of This Book

Wataru Nakayama
Chairman, ASME Japan Chapter

It is our pleasure to offer this book as a manifestation of our desire to do something useful under the auspices of the ASME Japan Chapter. The chapter was formed in 1986, originally as an informal gathering of ASME members residing in Japan. Since June 1990, it has been officially recognized as one of the ASME overseas chapters.

One of the Chapter's important functions is to provide an interface between ASME members of foreign origin and Japanese members. We decided to publish a newsletter printed in English to assist communication among members. I was privileged to serve as the editor for the first several issues of the ASME Japan Newsletter. For one of the issues, I invited Mr. Raymond Vonderau, then representative of Beloit Corporation in Tokyo, to contribute an article on his business experience in Japan. For another issue I solicited an article from Dr. Charles Knisely, then visiting lecturer at Kyoto University, on his experience in Japanese classrooms. Both articles received a great deal of attention even from outside ASME Japan, and I was flooded with requests from interested readers to publish their letters to the editor in subsequent issues. Unfortunately, the cost of printing did not permit me to add more pages to what is normally by a four-page newsletter. Besides, it was considered unwise to devote so much space in the newsletter to such serious topics as the culture shock experienced by foreign-born engineers and counterarguments from Japanese engineers. After consultation with other key members of ASME Japan, I decided to separate the issue of cultural interface from the newsletter, and create another channel to vent steam.

The first such channel was provided at the JSME spring meeting in April 1990, where the JSME International Exchange Committee and ASME Japan co-organized a panel discussion entitled "The Role of Engineering Societies in the Age of Internationalization." Professor Hideo Okamura, chairman of the JSME International Exchange Committee, and I cochaired the session, and we were privileged to include Professor Arthur E. Bergles, then ASME president-elect, and Professor Hideo Ohashi, then JSME vice president, among the panelists. The session was one of the most popular events at the 1990 JSME spring meeting, drawing more than 100 attendees. The success of the panel discussion urged us to consider recording what was discussed at the panel in some archival form, and it eventually led us to publish this book. I am confident, however, that this book is far richer in content than the proceedings of the panel session. In addition to the work by the panelists to enrich their manuscripts, other volunteers were invited to contribute articles. In order to add an objective aspect to the collection of articles, Mr. Raymond Vonderau took the trouble to produce a questionnaire to be given to Japanese companies and later reduced the response to statistical information.

The whole project of editing this book was entrusted initially to Drs. Hiroshi Honda and Kazuo Takaiwa, and later, as the project developed, to an enlarged editorial committee. All the committee members worked energetically to get our job done in a short time. Dr. Honda took care of the most burdensome task of all, corresponding with the authors, ASME Press, the companies, and the universities. The manuscripts for this book were reviewed by its editors and the editors appointed by ASME Press. An equally demanding task was the proofreading done by Mr. Daniel Day and other editors, whose volunteer service was indispensable to the publication of the book.

I sincerely thank all the authors for their valuable contributions, and join with them in their hope that this volume will serve a need of our time.

Activities of the ASME Japan Chapter

Ichiro Watanabe
Founding and Past Chairman, ASME Japan Chapter

At the time of publication, it is over five years since the ASME Japan Group was established at its first annual meeting, held on February 4, 1986, at the New Sanno Hotel in Tokyo, thanks to the earnest efforts of Stephen D. Lisse, P.E., then the ASME correspondent in addition to his duties as staff civil engineer for the U.S. Navy. The members of the ASME Japan Group were also members of JSME (the Japan Society of Mechanical Engineers), and it was clearly desirable that the activities of the ASME Japan Group not overlap with those of JSME. One of the main activities of JSME is to sponsor international conferences on mechanical engineering. Thus, after discussions by the members of the executive board of the group, it was concluded that the ASME Japan Group would place emphasis on the "short course seminars" given by lecturers from ASME headquarters, among others. Several seminars have been successfully held since then.

The ASME Japan Group was formally recognized as the Japan Chapter of ASME on June 6, 1990. We look forward to continuing and expanding our activities in the future.

About the Contributors

Michael W. Barnett is an American who is an assistant professor in the Department of Civil Engineering and Engineering Mechanics at McMaster University, Hamilton, Ontario. Prior to holding this position, he was a lecturer and a postdoctoral researcher studying time series analysis of environmental systems and artificial intelligence applied to control of water and wastewater treatment systems, in the Department of Environmental and Sanitary Engineering at Kyoto University. He originally came to Japan on a Ministry of Education scholarship and subsequently was hired as the first foreign-born faculty member in his department at Kyoto University. He holds a B.A. in psychology and an M.S. in environmental engineering, both from the University of Cincinnati, and a Ph.D. in environmental engineering from Rice University in Houston. He is a member of the International Association on Water Pollution Research and Control, the Water Pollution Control Federation, the American Society of Civil Engineers, and the Human Factors Society.

Arthur E. Bergles, an American, is 1990–91 president of the American Society of Mechanical Engineers and concurrently dean of engineering at Rensselaer Polytechnic Institute. A native of New York State, he received his B.S. and Ph.D. in mechanical engineering from MIT. He then served on the MIT staff and faculty from 1962 to 1969, as an associate professor and as chairman of the Engineering Projects Laboratory. He subsequently joined Georgia Institute of Technology as professor of mechanical engineering and remained until 1972, when he began an 11-year term as professor and chairman of mechanical engineering at Iowa State University. He joined the Rensselaer faculty in 1986, becoming the Clark and Crossan Professor of engineering, director of the Heat Transfer Laboratory, and then dean in September 1989. He has published more than 350 technical papers, books, and reports, and presented about 250 invited lectures in the United States and overseas. He received numerous awards from ASME and other professional societies, and is a fellow of ASME, ASEE, and AAAS.

Daniel K. Day, an American, is an English teacher and a free-lance translator (Japanese to English). His work includes weekly columns for the *Asahi Evening News*. He holds a B.S. in mechanical engineering from the

University of Colorado (Boulder). After working for Boeing Commercial Aircraft Corporation for $2\frac{1}{2}$ years, he moved to Sapporo, Japan, in December 1981.

Robert M. Deiters, an American who has lived in Japan since 1952, is a professor in the Electrical-Electronics Engineering Department of Sophia University in central Tokyo. He is a member of the Society of Jesus and an ordained Catholic priest. He holds degrees in theology from Sophia University and a doctorate in engineering (kogaku hakushi) from the University of Tokyo. At Sophia University he has been director of the Computer Center and dean of the Faculty of Science and Technology. At present he teaches and does research in computer networks and computer applications in engineering.

Jon R. Elmendorf, an American, is president of Westinghouse Energy Systems–Japan. Upon graduating from Duke University in 1968, he served in the U.S. Navy as a nuclear submarine officer. He joined Westinghouse Water Reactor Divisions in 1972 and held various positions in nuclear power plant services and construction and project management. During this period he received his MBA from the University of Pittsburgh. From 1981 through 1985 he was posted in Japan as managing director and later as executive vice president of Westinghouse Nuclear Japan. He returned to the United States in 1985, and in December of that year was named president of the O'Connor Combustor Corporation, a wholly owned subsidiary of Westinghouse in Fullerton, California. In February 1989, he returned to Tokyo to assume his present position.

Shuichi Fukuda is a Japanese who is a professor of systems engineering in the Management Engineering Department of the Tokyo Metropolitan Institute of Technology, and concurrently of the Institute of Industrial Science, University of Tokyo. He served as an associate professor of the Welding Research Institute at Osaka University and taught at the Department of Precision Machinery Engineering, University of Tokyo. He holds bachelor's, master's, and doctor's (kogaku hakushi) degrees in mechanical engineering from the University of Tokyo. His main interest is building intelligence into systems. He is a member of the trustees for several academic societies and a board member for the International Association for the Exchange of Students for Technical Experience.

Stephen A. Hann is an American who moved to Japan in 1989. He was raised in Virginia Beach, Virginia, and holds degrees in applied mechanics from Old Dominion University and in mechanical engineering from the University of Michigan. He and his wife, **Deborah A. Coleman Hann,** who was raised in Toledo, Ohio, established the Mechanical Simulation Corporation in Ann Arbor, Michigan, and moved to Japan in March 1989 in order to establish an office in Tokyo.

Charles R. Heidengren is an American consulting engineer. He holds a degree in civil engineering from the Cooper Union School of Engineering, with graduate studies in soil mechanics at Columbia University. He lived and worked in Japan from 1979 to 1989. He was manager for civil engineering in the engineering and construction division of Kawasaki Steel Corporation, headquartered in Tokyo. Prior to this, he spent five years as senior technical adviser and project manager for Pacific Consultants International. He has more than 26 years' experience in geotechnical and foundation engineering, with additional experience in project management and business development since 1979. During his career, he has made lecture presentations in the United States and Japan on transportation projects, foundation engineering, and relationships between owner, engineer, and contractor for international projects. He has published articles in *Civil Engineering* magazine, published by the American Society of Civil Engineers; the journal of the Japan Society of Civil Engineers (JSCE); and the journal of the Japanese Society of Soil Mechanics and Foundation Engineering (JSSMFE). He is a licensed professional engineer and was the founder and past president of the Japan Section of the American Society of Civil Engineers.

Hiroshi Honda is a Japanese who is an associate manager of the business planning department at Mitsui Engineering and Shipbuilding Company (MES). Prior to holding this position, he served as an associate manager of the corporate planning department and a chief research engineer at MES. He holds a bachelor's degree in mechanical engineering from Kyoto University, an M.S. degree in engineering mechanics from Pennsylvania State University, and a doctor's degree in engineering (*kogaku hakushi*) from Kyoto University. He is the author or coauthor of books, technical papers, and articles in the areas of machine elements, fracture mechanics, fatigue, strength design of structures and machinery, engineering education, intercultural topics, and management of R&D. He is a registered professional engineer in the states of Minnesota and Texas and a member of ASME,

JSME, ASTM, and SMSJ (Society of Materials Science, Japan). He serves as editorial vice chairman of the ASME Japan Chapter and is an alumni leader of the Penn State Club of Japan.

Peter E. D. Morgan is an American (formerly British) living in Thousand Oaks, California, where he works at Rockwell International Science Center. He has a Ph.D. in inorganic chemistry from Imperial College, London University, and is a fellow of the American Ceramic Society. He has visited Japan eight times, including a six-month stay at the Hitachi Corporation Superconductivity Center in Hitachi, Ibaraki prefecture. In his field of ceramics, he has more than 110 publications and patents, including five in Japanese periodicals.

Wataru Nakayama is a Japanese who is chairman of the ASME Japan Chapter and chairman of the Thermal Engineering Division of JSME, both for the 1990 term. Since 1970, Dr. Nakayama has been associated with the Mechanical Engineering Laboratory, Hitachi, Ltd. where he worked as a heat transfer specialist. He has been active in promotion of international exchanges of technical information, serving for organizing committees of various international conferences sponsored by JSME, ASME, and IEEE. Since 1989, he has been teaching and conducting research at the Department of Mechanical Engineering for Production, Tokyo Institute of Technology, as Hitachi Chair Professor, a newly created position to promote industry-academia interaction. He is an ASME fellow and IEEE senior member.

Hideo Ohashi, a Japanese, is a professor of mechanical engineering at the University of Tokyo. Prior to joining the faculty there, he worked on the design of gas turbines for five years at Ishikawajima Harima Heavy Industries (IHI). He holds bachelor of engineering and doctor of engineering (kogaku hakushi) degrees from the University of Tokyo and a Dr.-Ing. degree from Technische Universität Braunschweig, Germany. He specializes in the study of unsteady flow and two-phase flow phenomena in turbomachinery. He serves as president of the Turbomachinery Society of Japan and vice president of the Japan Society of Multi-Phase Flow, and served as vice president of the Japan Society of Mechanical Engineers (JSME) for 1989–90.

Kazuo Takaiwa is a Japanese who is a representative of ITK, Inc. (Internationale Technik und Konsulent Incorporation). He is a consultant for project engineering, construction, maintenance, training of company staff, and cross-cultural problems. He holds a bachelor's degree in aeronautics from Kyushu Imperial University and a doctor's degree (kogaku hakushi) in plant construction control systems from the University of Tokyo. He was a director, project manager, and construction manager for overseas projects and a pressure vessel factory manager at Chiyoda Corporation. He is the author or coauthor of six books in project control, staff training, and cross-cultural issues. He has authored over 30 technical papers and 60 articles. He is a member of ASME, AACE, PMI, JSME, and eight Japanese technological societies.

Craig Van Degrift is an American physicist in the Electricity Division of the National Institute of Standards and Technology (formerly, the National Bureau of Standards) carrying out research on the quantum Hall effect in support of the U.S. standard of electrical resistance. A native of Los Angeles, he received his B.S. in physics from Harvey Mudd College in 1966, an M.A. from the University of California at Irvine in 1967, and a Ph.D. from Irvine in 1974. After doing postdoctoral research on the high-frequency conductivity of metals, he joined the Heat Division of the National Bureau of Standards, Gaithersburg, Maryland. In 1985, he moved to the Electricity Division and built a new laboratory for the NBS (now NIST) quantum Hall effect research effort. During 1989, he was able to combine his professional physics activity with a long-term side interest in the Japanese language by visiting the Japanese Electrotechnical Laboratory for a year to participate in the Japanese quantum Hall effect research effort.

Raymond C. Vonderau is an American who is the regional vice president of the Tokyo office of Beloit Corporation (Beloit Nippon, Ltd.). He received a BSME degree from Purdue University in 1949 and joined Beloit in 1959. He has held various executive engineering positions for Beloit and its affiliate companies and has contributed to their success in the pulp and paper industry in North America. He is a member of the Technical Association of the Pulp and Paper Industry (TAPPI), ASME, and ASME Japan Chapter and is a registered professional engineer in the state of Wisconsin. He has been in his present position in Japan since 1986 and is currently a member of the Executive Committee for the ASME Japan Chapter.

Ichiro Watanabe is a Japanese professor emeritus at Keio University and consultant at the Institute of Science and Technology, Kanto Gakuin University. He was graduated from the University of Tokyo, and served as associate professor at the University of Tokyo as well as a member of the Aeronautical Research Institute of the University. He conducted investigations concerning flow patterns within centrifugal superchargers and high-altitude performance characteristics of aeronautical piston engines, among other things. He has served as professor at Keio University (1946–74), at Aoyama Gakuin University (1974–77), and at Kanto Gakuin University (1977–82), and as consultant at the Institute of Science and Technology, Kanto Gakuin University, since 1982. He received a doctor's degree in engineering (kagaku hakushi) from the University of Tokyo in 1947 and has been elected an honorary member of JSME, an honorary member of the Gas Turbine Society of Japan, and a fellow of ASME. He was the founding chairman of the ASME Japan Chapter and held the position of chairman for four years.

Dimitrios C. Xyloyiannis, a Greek, is working as a project manager and engineer for Sandoz Industrial Technology, Ltd., headquartered in Switzerland. He is involved in two projects in Japan. He is a professional marine engineer and holds a master's degree in mechanical engineering. He is a full member of the Greek Management Society and the Greek Operational Research Society (GORS). He was a lecturer in technical seminars offered by GORS until 1989, and is a member of ASME in the United States and Japan.

Ghassem Zarbi, an Iranian, is a guest researcher studying hydraulic power control at Sophia University in Tokyo. He holds B.Sc., M.Sc., and M.Phil. degrees in mechanical engineering and manufacturing technology from universities in London and has submitted a thesis for a doctor's degree (kogaku hakushi) in mechanical engineering from Sophia University. He worked for a Japanese manufacturer of machine elements as technical adviser for two years and was a technical adviser and managing director for two years for two different American and Japanese CAD/CAM software developers before establishing his own computer training center. He has authored 12 technical papers. He serves as secretary of ASME Japan Chapter.

Glossary of Japanese Words

ba	Place and frame for one's life and activity. (48)
bucho	General manager. (20)
chiho	Region. (66)
chodendo	Superconductivity. (114)
daigaku	University. (45)
dan-dan	Step by step. (117)
deru kugi wa utareru	"The nail that sticks out is hammered down"; proverb describing Japanese wariness of unusual personality or individuality, or of expression of "Hold down outstanding characters and/or talent that may stand out from the group." (7)
fusuma	Japanese paper door. (85)
gaijin	Foreigner. (69)
Gonin-gumi	Five-person team through which families watched each other's behavior, under the social system that evolved during the Edo period. (30)
haragei	The art (*gei*) of the belly (*hara*), where the "belly" signifies one's heart, what one is really thinking. The art is in transmitting one's intention without putting it directly into words. (62)
harakiri	Committing suicide by cutting the belly with a sword. (31)
hara o yomu	To read another's intention. (62)
hiragana	Kind of kana used for genuine Japanese words. (65)
honne	One's true intention. (62)
Honshu	Japan's main island. (148)
kacho	Manager. (21)
kaizen	Improvement, implying many small, never ending steps. (117)

kakaricho	Assistant manager. (21)
kamikaze	"The wind of the gods," applied to a typhoon that stopped a Mongolian force attempting to invade Japan in the thirteenth century. (87)
kana	Japanese syllabic character; closest counterpart of alphabetic characters in English. (64)
kanji	Chinese character, used in the Japanese language to signify the meaning of a word, not its sound. (64)
katakana	A kind of kana used for words of western origin. (66)
ken	Prefecture. (111)
kogaku hakushi	Doctor's degree in engineering conferred by Japanese universities, which certifies that the person has made substantial achievement in engineering research and has passed a qualifying examination. (9)
kogakushi	Bachelor's degree in engineering conferred by Japanese universities.
kogaku shushi	Master's degree in engineering conferred by Japanese universities.
kokoro	Soul or spirit. (118)
koohai	Younger person in "vertical" human relations. (49)
Monbusho	Ministry of Culture and Education. (130)
nanushi	Village headman in feudal periods in Japan. (30)
nemawashi	Literally, "root binding," or gaining a consensus of the necessary people. (35)
Nihongo	The Japanese language. (65)
nominucation	A slang combination of the Japanese word *nomi,* "drinking," and the English word *communication.* (35)
romaji	Roman alphabet. (63)
sake	Rice wine. (117)
sakura	Cherry blossom. (118)
samurai	Social rank in feudal periods in Japan equivalent to knight. (30)

sempai	Older person in "vertical" human relations. (49)
shi	City. (111)
shoji	Sliding paper door. (54)
shosha	Commodity trading company. (4)
tabi	Traditional Japanese-style socks. (93)
tatami	Woven straw mats regularly used in Japanese houses. (54)
tatemae	The way something is supposed to be, in contrast with *honne*. (62)
uchi-soto	Insider (uchi) and outsider (soto); us and them; or your group and competing group. (117)
wa	Peace, harmony, and cooperation. (7)
yakuin	Executive, such as director, managing director, vice president, president, or chairman. (20)

Index